Recent Advances in Lifetime and Reliability Models

Authored by:

Gauss M. Cordeiro, Rodrigo B. Silva & Abraão D. C. Nascimento

Recent Advances in Lifetime and Reliability Models

Authors: Gauss M. Cordeiro, Rodrigo B. Silva & Abraão D. C. Nascimento

eISBN (Online): 978-1-68108-345-2

ISBN (Print): 978-1-68108-346-9

need for a court order if at any point you breach any terms of this License Agreement. In no event will any delay or failure by Bentham Science Publishers in enforcing your compliance with this License Agreement constitute a waiver of any of its rights.

3. You acknowledge that you have read this License Agreement, and agree to be bound by its terms and conditions. To the extent that any other terms and conditions presented on any website of Bentham Science Publishers conflict with, or are inconsistent with, the terms and conditions set out in this License Agreement, you acknowledge that the terms and conditions set out in this License Agreement shall prevail.

Bentham Science Publishers Ltd.
Executive Suite Y - 2
PO Box 7917, Saif Zone
Sharjah, U.A.E.
Email: subscriptions@benthamscience.net

Contents

Foreword

This book is, as far as I have gathered, the first book of its kind. The authors should be commended for spending countless hours researching the literature and explaining in details the connections between many different distributions published during the past decades. I believe there will be so many grateful researchers and readers who will have a broad perspective of all the interesting distributions presented in this manuscript. This book can serve as a foundation for those who are seriously interested in doing research in the field of distribution theory. The content of the book deals with a comprehensive treatment of methods for lifetime models, which has many practical applications in various fields. It seems to me, if one wants to do literature review of the published work in this area, all one needs to do is get hold of a copy of this book. In my opinion, this book is destined to be an extremely important source for motivating the younger researches in the field of distribution theory, in particular lifetime models. This captivated book represents a complete account of important distributions, their properties and their applications in various fields of applied sciences. I am sure that this book will serve as a unique and excellent source of information in the overall field of statistics and probability for many years to come. I admire the effort of the authors to come up with such a fantastic work.

G.G. Hamedani
Marquette University
Milwaukee, Wisconsin
USA

Preface

The theory of distributions with support in the positive real numbers has grown and matured in the last two decades, becoming one of the main statistical tools for the analysis of lifetime (survival) data. In fact, in many ways, lifetime distributions are the common language of survival dialogue because the framework subsumes many statistical properties of interest, such as reliability, entropy and maximum likelihood.

This book provides a comprehensive account of models and methods for lifetime models. Building from primary definitions such as density and hazard rate functions, the book presents the distribution theory in survival analysis. This framework covers classical methods, such as the exponentiated method, and also the most recent developments in lifetime distributions, such as the beta family and compounding models. Additionally, there is a detailed discussion of mathematical and statistical properties of each family, such as mixture representations, asymptotes, some types of moments, order statistics, quantile and generating functions and estimation. There is also a brief exploration of regression models for the beta generalized family of distributions. Throughout the text, we focus not only on the theoretical arguments but also on issues that arise in implementing the statistical methods in practice. The most recent parametric models in lifetime data analysis are covered without concentrating exclusively on any specific field of application, and most of the examples are drawn from engineering and biomedical sciences. It is important to emphasize that even with omission of some models, the great amount of models available has forced us to be very selective for inclusion in this work. To keep the book at a reasonable length we have had to omit or merely outline certain models that might have been included.

To help readers, lists of notation, terminology, and some probability distributions are given at the beginning of the book. All notational conventions are the same or very similar to the articles from which the models are based. Readers are assumed to have a good knowledge in advanced calculus. A course in real analysis is also recommended. If this book is used with a statistics textbook that does not include probability theory, then knowledge in probability theory is required.

The main five generators of new distributions are grouped into seven sections corresponding to those to which they give names. Chapter 1 contains introductory material with mathematical and statistical background for understanding this book. Chapter 2 deals with the exponentiated method. Explicit expressions for the quantile function, ordinary and incomplete moments, probability weighted moments, cumulants and generating functions are presented for the exponentiated-G family. Chapter 3 discusses the procedure that generates what we call the beta generalized family. Further, useful expansions and several statistical properties are presented. Chapter 4 provides theoretical essays about five special models in the beta family. For each model, its cumulative, density and hazard rate functions have explicit forms and important linear representations, which can be used to obtain some mathematical properties. Two applications are performed in order to illustrate the flexibility of the densities under discussion. Chapter 5 introduces the Kumaraswamy generalized family. In addition, several structural properties are presented and discussed for this family. Among them, useful expansions, quantile and generating functions, moments and mean deviations. Additionally, estimation and generation procedures are investigated. Chapter 6 presents three special cases of the Kumaraswamy generalized family. Some mathematical properties are provided such as the moments and generating function. Useful expansions for the density function and some special cases are presented. Chapter 7 discusses the gamma generalized family proposed by Zografos and Balakrishnan (2009). Several mathematical properties are provided such as expansions for the density and cumulative functions, quantile function, moments, generating function and entropies. A bivariate generalization is presented. Chapter 8 introduces a family of models defined by compounding two (a continuous and other discrete) distributions. We provide important mathematical properties such as moments and order statistics. We discuss the estimation of the model parameters by maximum likelihood and prove empirically the potentiality of the family by means of two applications to real data.

Readership

We hope that this book inspires students that make extensive use of observational data, including finance, medicine, biology, sociology, education, psychology, engineering and climatology. Further, we hope that our readers come to regard this book as a reliable source of information and we gladly welcome all efforts to bring any remaining errors to our attention.

Acknowledgements

Authors would like to thank the financial assistance received from *Conselho Nacional de Desenvolvimento Científico e Tecnológico* - CNPQ. Our indebtedness also extends to Statistic Departments at Federal University of Pernambuco and Federal University of Paraíba by providing some facilities.

Conflict of Interest

The authors endorse that the Book content has no conflict of interest.

GAUSS M. CORDEIRO
UNIVERSIDADE FEDERAL DE PERNAMBUCO, BRAZIL
RODRIGO B. SILVA
UNIVERSIDADE FEDERAL DA PARAÍBA, BRAZIL
ABRAÃO D.C. NASCIMENTO
UNIVERSIDADE FEDERAL DE PERNAMBUCO, BRAZIL

Nomenclature

ACRONYMS

AIC	Akaike Information Criterion
BBIII	the beta Burr III distribution
BBS	the beta Birnbaum-Saunders distribution
BBXII	the beta Burr XII distribution
BC	Bonferroni curve
BE	the beta exponential distribution
BFr	the beta Fréchet distribution
BFGS	the Broyden-Fletcher-Goldfarb-Shanno optimization method
BG	the beta-G family of distributions
BGE	the beta generalized exponential distribution
BHC	the beta half-Cauchy distribution
BI	Bonferroni index
BIC	Bayesian Information Criterion
BLa	the beta Laplace distribution
BLN	the beta lognormal distribution
BMW	the beta modified Weibull distribution
BN	the beta normal distribution
BPa	the beta Pareto distribution
BSL	the beta standard logistic distribution
BSPS	the Birbaum-Saunders power series distributions
BW	the beta Weibull distribution
BWG	the beta Weibull geometric distribution
BXIIPS	the Burr XII power series distributions
CAIC	Consistent Akaike Information Criterion
cdf	cumulative distribution function
cf	characteristic function
cgf	cumulant generating function
chf	cumulative hazard function

cmgf	central moment generating function
EE	the exponentiated exponential distribution
EFr	the exponentiated Fréchet distribution
EG	the exponentiated gamma distribution
EGu	the exponentiated Gumbel distribution
EM	expectation maximization
EV	the extreme value distribution
exp-G	the exponentiated-G class of distributions
EW	the exponentiated Weibull distribution
EWPS	the extended Weibull power series distributions
GE	the generalized exponential distribution
GI	Gini index
GoF	Goodness-of-Fit
GW	the generalized Weibull distribution
hrf	hazard rate function
KS	Kolmogorov Smirnov Statistic
LBBS	the log-beta Birnbaum-Saunders distribution
LBW	the log-beta Weibull distribution
LC	Lorenz curve
log-BW	the log-beta-Weibull distribution
LW	the log-Weibull distribution
mgf	moment generating function
MixEW	the mixture of Weibull distributions
MLE	maximum likelihood estimate
mrlf	mean residual life function
pdf	probability density function
PWM	probability weighted moment
rhrf	reversed hazard rate function
sf	survival function
1-MEW	the first modified Weibull distribution
2-MEW	the second modified Weibull distribution
3-MEW	the third modified Weibull distribution

NOTATIONS

$h_{\bullet}(\cdot),\ h(\cdot),\ h(\cdot;\cdot)$	hrf
$S_{\bullet}(\cdot),\ S(\cdot)$	sf
$H_{\bullet}(\cdot),\ H(\cdot)$	chf
$m(\cdot)$	mrlf
T, Y, X, T^{*}	Random variables (specifically, T describes failure time)

$\boldsymbol{\theta}, \boldsymbol{\beta}$	parameters vectors
$L(\boldsymbol{\theta})$	likelihood function
$\ell(\boldsymbol{\theta})$	log-likelihood function
$\boldsymbol{K}(\boldsymbol{\theta})$	Fisher's information matrix
$\boldsymbol{I}(\boldsymbol{\theta})$	observed information matrix
$\overset{a}{\sim}$	to be asymptotically distributed
$\mathcal{U}$	index sets for uncensored data
$\mathcal{C}$	index sets for censored data
$k(\cdot, \cdot)$	link function
$h_0(\cdot)$	baseline hrf
$S_0(\cdot)$	baseline sf
$\Gamma(\cdot)$	gamma function
$\psi^{(n)}(\cdot),\ \psi(\cdot)$	polygamma and digamma functions
γ, C	Euler-Mascheroni constant
$\Gamma(\cdot, \cdot),\ \gamma(\cdot, \cdot)$	incomplete gamma functions
$\Gamma_1(\cdot, \cdot),\ \gamma_1(\cdot, \cdot)$	regularized incomplete gamma functions
$M(\cdot, \cdot, \cdot)$	Kummer's first kind confluent hypergeometric function
$U(\cdot, \cdot, \cdot)$	Kummer's second kind confluent hypergeometric function
$N(\mu, \sigma^2)$	the normal (Gaussian) distribution with mean μ and variance σ^2
$\Phi(\cdot)$	standard normal cdf
$\phi(\cdot)$	standard normal pdf
$\mathrm{erf}(\cdot),\ \mathrm{erfc}(\cdot)$	error function amd its counterpart
$B(\cdot, \cdot),\ B_x(\cdot, \cdot)$	beta and incomplete beta functions
$I_x(\cdot, \cdot)$	beta cdf
$\eta(\cdot)$	Riemann's zeta function
$F(\cdot, \cdot; \cdot, \cdot),\ {}_2F_1(\cdot, \cdot; \cdot, \cdot)$	confluent hypergeometric function
$(z)_n$	Pochhammer polynomial
$G_{p,q}^{m,n}(\cdot)$	the Meijer G-function
$J_\tau(x)$	Bessel function of the first kind
$F_A^{(n)}(\cdot; \cdot; \cdot; \cdot)$	Lauricella function of type A
$F_{C:D}^{A:B}(\cdot; \cdot; \cdot)$	generalized Kampé de Fériet function
${}_p\Psi_q$	complex parameter Wright generalized hypergeometric function
$(\Omega, \mathcal{F}, P)$	probability space
$\mu'_k,\ \mu_k,\ \mu'_{(k)}$	kth moment on zero, kth central moment and descending factorial moment

CV	coefficient of variation
γ_1, γ_2	skewness and kurtosis
$M_X(t)$, $K_X(t)$	mgf and cgf
κ_k	kth cumulant
$\phi_X(t)$	cf
$\tau_{k,l}$	PWM
$B_F(\cdot)$, $L_F(\cdot)$	BC and LC
B, G	BI and GI
$H_{\mathrm{R}}^{\beta}(X)$, $H_{\mathrm{S}}(X)$	Rényi (with order β) and Shannon entropies of X
R	Reliability
$X_{i:n}$, $F_{X_{i:n}}(x)$, $f_{X_{i:n}}(x)$	ith order statistic and its cdf and pdf
$D(\cdot,\cdot)$	distance measure
A^*	modified Anderson-Darling statistic
W^*	modified Cramér-von Mises statistic

Chapter 1

Introduction

Abstract: This chapter presents mathematical and statistical background for understanding this book. Some results and formulae presented in this chapter are revisited in the next chapters. Initially, important survival analysis concepts are defined and issues with respect to inference and statistical methods. Subsequently, several special functions are presented. The chapter ends with some discussions on statistical elements that will be used in the rest of the book.

Keywords:: Censoring data; Inference; Mathematical functions; Statistical functions; Survival functions; Survival regression.

Lifetime statistical analysis is an important subject in applied areas like biomedical science, engineering, reliability, social sciences and several others. Typically, the term *lifetime* or *failure* can have different interpretations. According to Lai (2011) [1], it can represent:

- the human life age [2],

- the time operation of an equipment to fail [3],

- the survival time of a patient with a serious disease from the date of diagnosis [4],

- the time to first recurrence of a tumor (*i.e.*, length of remission) after initial treatment or

- the duration of a social event such as marriage [2].

In above practical occurrences, a failure can not be computed either by an imposed contextual criterion or due to a stochastic censoring. For instance, it can be seen whereas a patient does not die during a clinical treatment period or if he (or she) leaves the trial process. Thus, the proposal of analysis methods that incorporate censoring as well as procedures for failure time data has been sought. *Survival analysis* is the set of statistical procedures able to describe time-to-event censored data. An important step to deal with survival data consists at proposing more flexible models, which furnish a good representation for both nature of data and the shape of its empirical distribution. This book presents a comprehensive mathematical treatment about the main classes of distributions for describing lifetime data.

New distributions often result from a modification of a baseline random variable X by (i) linear transformation, (ii) power transformation (*e.g.* the Weibull is obtained from the exponential), (iii) non-linear transformation (*e.g.* the log-normal from the normal), (iv) log transformation (*e.g.* the log Weibull, also known as the type 1 extreme value distribution), and (v) inverse transformation (*e.g.* the inverse Weibull and inverse gamma models). In what follows, we present two simple transformations for generating new models.

POWER TRANSFORMATION

Consider $G(x)$ be the original cumulative distribution function (cdf) and $F(x)$ be the cdf of a new ageing distribution derived from $Y \sim G$ by exponentiating as follows:

$\boxed{F(t) = G(t)^{\alpha}}$: Using such power transformation, one can deduce the generalized modified Weibull distribution proposed by Carrasco *et al.* (2008) [5], the exponentiated Erlang distribution by Lai (2010) [6] and the exponentiated Weibull by Mudholkar and Srivastava (1993) [7].

$\boxed{F(t) = 1 - \{1 - G(t)\}^{\beta}}$: The Lomax model can be formulated from the Pareto distribution in this way.

MIXTURE OF DISTRIBUTIONS

New models are often obtained from mixtures of two or more distributions. Let π be the mixing proportion of two cdfs $F_1(t)$ and $F_2(t)$. The cdf $F(t)$ resulting from mixture between the two cdfs is

$$\boxed{F(t) = \pi F_1(t) + (1 - \pi) F_2(t).}$$

In this book, we present the formalisms of five new classes or generators of distributions, which have been used for describing high complexity data; in particular, for the survival analysis context. The background for understanding class concepts and associated applications is presented in the rest of this chapter.

1.1.   PRIMARY DEFINITIONS

Let $T \geq 0$ denote the lifetime random variable having $f_T(t)$ and $F_T(t) = \int_0^t f_T(x)\, \mathrm{d}x$ as probability density function (pdf) and cdf, respectively. In this book, we consider that T is an absolutely continuous random variable (for a discussion on discrete lifetime models, see Lawless (1982, p. 10) [8]). In this case, $S(t) = \bar{F}_T(t) = 1 - F_T(t) = \int_t^\infty f_T(x)\, \mathrm{d}x$ is defined as *reliability* or *survival function* (sf). It is obvious that $S(t)$ is a monotone decreasing function with $S(0) = 1$ and $S(\infty) = \lim_{t \to \infty} S(t) = 0$. The pdf can be expressed in terms of $S(t)$ as

$$f_T(t) = \lim_{\Delta t \to 0^+} \frac{P(t \leq T < t + \Delta t)}{\Delta t} = \frac{\mathrm{d}F_T(t)}{\mathrm{d}t} = -\frac{\mathrm{d}S(t)}{\mathrm{d}t}.$$

An important concept is the *hazard rate function* (hrf) defined as

$$h_T(t) = \lim_{\Delta t \to 0^+} \frac{P(t \leq T < t + \Delta t \mid T \geq t)}{\Delta t} = \frac{f_T(t)}{1 - F_T(t)} = \frac{f_T(t)}{S(t)} \quad (1.1)$$

and, therefore, "$h_T(t)\,\Delta t$" returns the probability of failure in $(t, t + \Delta t]$ given the "unit" has survived until time t. The hrf is also referred to as the *risk* or *mortality rate*. The functions $\bar{F}_T(\cdot)$ and $h_T(\cdot)$ are also called as *ageing* measures. There are several other measures of ageing, but we discuss the hazard and survival functions because they are the major ones in reliability practice. Further, we have

$$h_T(t) = -\frac{\mathrm{d}S(t)/\mathrm{d}t}{S(t)} = -\frac{\mathrm{d}\log[S(t)]}{\mathrm{d}t}$$

and, therefore, the *cumulative hazard function* (chf), $H_T(t)$, is

$$H_T(t) = \int_0^t h_T(u)\, \mathrm{d}u = -\log[S(t)]$$

$$\Leftrightarrow S(t) = \exp\left[-H_T(t)\right] = \exp\left[-\int_0^t h_T(u)\, \mathrm{d}u\right].$$

Thus, the pdf of T can be expressed from (1.1) as

$$f_T(t) = h_T(t) \exp\left[-\int_0^t h_T(u)\, \mathrm{d}u\right].$$

Moreover, from probability basic results for non-negative random variables, one has that $\mathrm{E}(T) = \int_0^\infty S(t)\, \mathrm{d}t$, *i.e.*, the mean survival time is the total area under the survival curve. From the last identity, one has that the applied pattern is "the greater risk is associated to the shorter mean survival time". Other important concept is the *mean residual life function* (mrlf). The mrlf (or the function which measures the remaining life expectancy at age t) is defined by

$$m(t) = \mathrm{E}(X - t \mid X > t) = \frac{\int_t^\infty S(x)\, dx}{S(t)}.$$

In order to illustrate the possible behaviours of $h_T(t)$, consider that T follows the generalized exponential geometric (GEG) distribution with parameter vector $\boldsymbol{\theta} = [\alpha, \beta, p]^\top$ and hrf given by

$$h_T(x; \boldsymbol{\theta}) = \frac{\alpha\,\beta\,(1-p)\,e^{-\beta x}\,(1 - e^{-\beta x})^{\alpha-1}\,(1 - p\,e^{-\beta x})^{\alpha-1}}{(1 - p\,e^{-\beta x})^\alpha - (1 - e^{-\beta x})^\alpha}, \tag{1.2}$$

which was proposed by Silva *et al.* (2010) [9] and will be detailed in Section 8.1. We plot some hazard curves in Figure 1.1. This case is very didactic because they proved that the behavior of (1.2) can be defined as in the Table 1.1.

Table 1.1: Parameter intervals with the corresponding shapes of the failure rate.

p \ α	$(0,1]$	$(1,\infty)$
$\left(0, \frac{\alpha-1}{\alpha+1}\right]$	Decreasing	Increasing
$\left(\frac{\alpha-1}{\alpha+1}, 1\right)$	Decreasing	Upside-down bathtub

1.2. CENSORING KINDS

In this section, we discuss briefly two types of censoring for failure data. A detailed discussion can be found in Tableman and Kim (2005) [10].

1.2.1. First Censoring

Consider we want to assess "n" items. Assume also that all of them will be put in test and record their times to failure. However, we will not wish to wait beyond a pre-specified time t_c. The number of observed failure times, N, is used as a random quantity and, then, $N = 0, 1, 2, \ldots$.

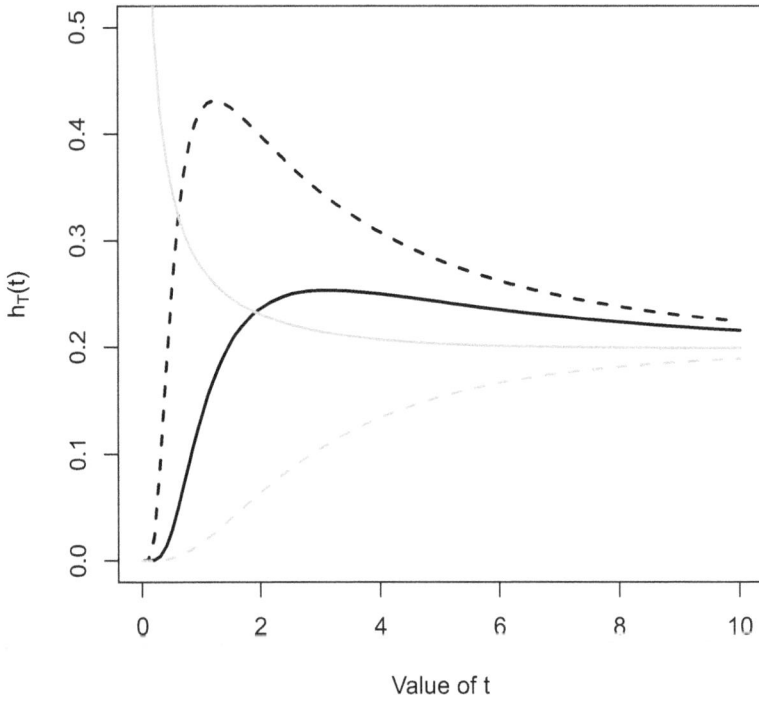

Figure 1.1: Decreasing (gray solid curve), increasing (gray dashed curve) and upside-down bathtub hazard function (solid and dashed black curves).

Let $T_1, \ldots, T_n$ be a random sample drawn from T, which describes the failure time and has cdf given by $F_T(t)$. Note that we observe from the above context $\boldsymbol{y} = (Y_1, \ldots, Y_n)^\top$ such that

$$Y_i = \min(T_i, t_c) = \begin{cases} T_i, & T_i \leq t_c \\ t_c, & T_i > t_c. \end{cases}$$

Now, we have that $\boldsymbol{y}$ is a random sample from Y with cdf

$$F_Y(y) = P(T_i \leq t_c) \, F_T(y) + \underbrace{P(T_i > t_c)}_{S(t_c)} \, F_{T^*}(y),$$

where T^* is a degenerate random variable with $P(T^* = t_c) = 1$ and $F_{T^*}(t) = \mathbb{I}(t \geq t_c)$, where $\mathbb{I}(t \geq t_c)$ is one for $t \geq t_c$ and zero for $t < t_c$. In order to define the associated likelihood, consider the auxiliary variable $\delta_i = \mathbb{I}(t_i \leq t_c)$, where t_i is an outcome of T_i. Thus, the likelihood and log-likelihood functions for the parameter vector $\boldsymbol{\theta}$ from n independent and identically distributed (iid) pairs (y_i, δ_i) can be defined, respectively, as

$$L(\boldsymbol{\theta}) = L(\boldsymbol{\theta}; \{y_1, \ldots, y_n\}) = \prod_{i=1}^{n} [f(y_i)]^{\delta_i} [S(t_c)]^{1-\delta_i}$$

and

$$\ell(\boldsymbol{\theta}) = \sum_{i=1}^{n} \{ \delta_i \log[f(y_i)] + (1 - \delta_i) \log[S(t_c)] \}.$$

1.2.2. Second Censoring

A second kind of censoring is given as follows: Consider we want to observe failure times until a pre-specified percentage, say $r/n\,100\,\%$, of n items have failed. It is the same that the r smallest elements from a n-points random sample from T (a model for failure data), $T_1, \ldots, T_n$. Then, the vector of random variables under study is $\boldsymbol{y} = (Y_1, \ldots, Y_n)^\top$ such that

$$Y_i = \begin{cases} T_{i:n}, & i \leq r, \\ T_{r:n}, & i > r, \end{cases}$$

where $T_{i:n}$ is the ith order statistic of $T_1, \ldots, T_n$ with cdf and pdf

$$F_{i:n}(x) = \sum_{k=i}^{n} \binom{n}{k} F(x)^k [1 - F_X(x)]^{n-k}$$

and

$$f_{i:n}(x) = \frac{n!}{(i-1)!(n-i)!} f_X(x) \, F_X(x)^{i-1} \left[1 - F_X(x)\right]^{n-i},$$

respectively. In this case, the likelihood function is

$$L(\boldsymbol{\theta}) = \frac{n!}{(n-r)!} \left[f_T(t_{1:n}) \cdots f_T(t_{r:n})\right] \times S(t_{r:n})^{n-r},$$

where $t_{i:n}$ is an outcome of $T_{i:n}$.

1.2.3. Parametric Estimation in Failure Data

Let $Y_i = (T_i, C_i)$ such that T_i and C_i represent failure time and censoring, respectively. In general, the log-likelihood function is

$$\ell(\boldsymbol{\theta}) = \sum_{i \in \mathcal{U}} \log f_T(y_i | \boldsymbol{\theta}) + \sum_{i \in \mathcal{C}} \log S(y_i | \boldsymbol{\theta}), \tag{1.3}$$

where $\mathcal{U}$ and $\mathcal{C}$ represent index sets on which uncensored and censored data are observed. Based on equation (1.3), the maximum likelihood estimator (MLE) of $\boldsymbol{\theta}$ is defined by

$$\widehat{\boldsymbol{\theta}} = \arg\max_{\boldsymbol{\theta} \in \Theta} \{\ell(\boldsymbol{\theta})\},$$

where Θ represents the parametric space. Henceforward, we use MLE for both ML estimate and estimator. Assuming that the identity

$$\frac{\mathrm{d}}{\mathrm{d}\boldsymbol{\theta}} \int_{-\infty}^{\infty} f(x; \boldsymbol{\theta}) \mathrm{d}x = \int_{-\infty}^{\infty} \frac{\mathrm{d}}{\mathrm{d}\boldsymbol{\theta}} f(x; \boldsymbol{\theta}) \mathrm{d}x$$

holds, the expected information matrix (also known as Fisher's Information matrix) is defined as

$$\boldsymbol{K}(\boldsymbol{\theta}) = \mathrm{E}\left\{\frac{\partial \ell(\boldsymbol{\theta})}{\partial \boldsymbol{\theta}} \left[\frac{\partial \ell(\boldsymbol{\theta})}{\partial \boldsymbol{\theta}}\right]^{\top}\right\} = \mathrm{E}\left[-\frac{\partial^2 \ell(\boldsymbol{\theta})}{\partial \boldsymbol{\theta} \partial \boldsymbol{\theta}^{\top}}\right].$$

Under standard regularity conditions, the MLE $\widehat{\boldsymbol{\theta}}$ has the following asymptotic distribution

$$\widehat{\boldsymbol{\theta}} \overset{a}{\sim} N_p(\boldsymbol{\theta}, \boldsymbol{K}(\boldsymbol{\theta})^{-1}), \tag{1.4}$$

where $\overset{a}{\sim}$ represents "to be asymptotically distributed" and $\boldsymbol{K}(\boldsymbol{\theta})^{-1}$ is the large sample covariance matrix of $\widehat{\boldsymbol{\theta}}$. In practice, a closed-form for $\boldsymbol{K}(\boldsymbol{\theta})$ is difficult or, sometimes, an intractable task. Thus, the observed information matrix defined by $\boldsymbol{J}(\boldsymbol{\theta}) = -\partial^2 \ell(\boldsymbol{\theta})/\partial \boldsymbol{\theta} \partial \boldsymbol{\theta}^{\top}$ is used as an estimator of $\boldsymbol{K}(\boldsymbol{\theta})$.

Based on $K(\boldsymbol{\theta})$ or $J(\boldsymbol{\theta})$, we may adopt the likelihood ratio (LR) statistic to verify if the fit due to the new distribution outperforms meaningfully the fits of baseline laws in a data set. Let $\boldsymbol{\theta} = (\boldsymbol{\theta}_1^\top, \boldsymbol{\theta}_2^\top)^\top$ be the parameter vector of a proposed model, where $\boldsymbol{\theta}_1$ is a subset of parameters of interest (which can also be understood as additional parameters of a new model with respect to a standard distribution) and $\boldsymbol{\theta}_2$ is a subset of the remaining parameters. The LR statistic for testing the null hypothesis $\mathcal{H}_0 : \boldsymbol{\theta}_1 = \boldsymbol{\theta}_1^{(0)}$ versus the alternative hypothesis $\mathcal{H}_1 : \boldsymbol{\theta}_1 \neq \boldsymbol{\theta}_1^{(0)}$ is given by $w = 2\{\ell(\widehat{\boldsymbol{\theta}}) - \ell(\widetilde{\boldsymbol{\theta}})\}$, where $\widehat{\boldsymbol{\theta}}$ and $\widetilde{\boldsymbol{\theta}}$ are the MLEs based on observed sample of size n under the alternative and null hypotheses, respectively. The statistic w is asymptotically (as $n \to \infty$) distributed as χ_k^2, where k is the dimension of the subset $\boldsymbol{\theta}_1$ of interest.

For censoring data, it is important to determinate the sf, $S(y_i) = S(y_i|\boldsymbol{\theta})$, from data, and its estimated value $\widehat{S}(y_i)$. Let Y_i have mean μ and variance σ^2 and consider that $S'(y_i) < \infty$. By taking the Taylor expansion of $S(y_i)$ around μ_i and suppressing the high-order terms, we have

$$S(y_i) \approx S(\mu_i) + (Y_i - \mu_i)\, S'(\mu_i).$$

Thus, $\mathrm{E}(S(Y_i)) \approx S(\mu_i)$ and $\mathrm{Var}(\mathrm{S}(\mathrm{Y_i})) \approx [S'(\mu_i)]^2\, \sigma^2$. From equation (1.4) and applying the delta method [10], we obtain

$$\widehat{S}(Y_i) \overset{a}{\sim} N(S(\mu), [S'(\mu)]^2\, \sigma^2).$$

1.3. SURVIVAL REGRESSION MODEL

Let T be a model to describe failure time and $\boldsymbol{x} = (x_1, \ldots, x_p)^\top$ be a vector of covariates. We are interested in mapping a relation between $\boldsymbol{x}$ and a survival type-moment of T. In general terms, the introduction of linear predictor into the failure model is made by means of its hrf:

$$h(t; \boldsymbol{x}) = h_0(t)\, k(\boldsymbol{x}; \boldsymbol{\beta}), \qquad (1.5)$$

where $\boldsymbol{\beta} = [\beta_1, \ldots, \beta_p]^\top$ is the parameter vector, $k(\cdot; \cdot)$ is the *link function* and $h_0(t)$ is called the *baseline hazard function*. It is known from the regression theory that other issues besides to present modelling schemes (such as (a) to define hypothesis tests for checking if any subsets of $m < p$ covariates from (1.5), $\boldsymbol{x}$ is meaningful in explaining survival time as well as (b) to describe how to make a diagnostic analysis for this type of modelling), are important. However, in this section, we consider only models to survival data since the book focuses on new generators of continuous distributions.

In what follows, we present briefly two models to that end: *Cox proportional hazards model* and *accelerated failure time model*.

1.3.1.   Cox Proportional Hazards Model

In this case, the link function (1.5) is $k(\boldsymbol{x}; \boldsymbol{\beta}) = \exp(\boldsymbol{x}^\top \boldsymbol{\beta})$ and the hrf is

$$h(t; \boldsymbol{x}) = h_0(t) \exp(\boldsymbol{x}^\top \boldsymbol{\beta}). \tag{1.6}$$

For fixed two p-dimensional points $\boldsymbol{x}$ and $\boldsymbol{y}$, the following hazards ratio is constant regards to t:

$$\frac{h(t; \boldsymbol{x})}{h(t; \boldsymbol{y})} = \exp[(\boldsymbol{x} - \boldsymbol{y})^\top \boldsymbol{\beta}].$$

This fact indicates the proportional hazards property. In this case, the survivor function of T is

$$S(t; \boldsymbol{x}) = \exp\left(-\int_0^t h(u; \boldsymbol{x}) \, du\right) = \exp\left(-\exp(\boldsymbol{x}^\top \boldsymbol{\beta}) \int_0^t h_0(u) \, du\right)$$
$$= [S_0(t)]^{\exp(\boldsymbol{x}^\top \boldsymbol{\beta})},$$

where $S_0(t) = \exp\left(-\int_0^t h_0(u) \, du\right)$.

Now, consider a construction of a regression model for $T \sim W(\alpha, \lambda)$ having pdf given by (for $t > 0$)

$$f_T(t) = \lambda \, \alpha \, (\lambda \, t)^{\alpha-1} \, e^{-(\lambda t)^\alpha},$$

where $\alpha > 0$ and $\lambda > 0$ represent the shape and scale parameters, respectively. In this case, $S(t) = e^{-(\lambda t)^\alpha}$ and $h(t) = \lambda \alpha (\lambda t)^{\alpha-1}$. Thus, the relationship between $H(t)$ and $S(t)$ is given by

$$\log[H(t)] = \log\{-\log[S(t)]\} = \alpha \left[\log(\lambda) + \log(t) \right]$$

and, therefore,

$$\log(t) = -\log(\lambda) + \sigma \log\{-\log[S(t)]\},$$

where $\sigma = \alpha^{-1}$. The last identity reveals that the term $\log(t)$ can be explained by a straight line with slope $\sigma = \alpha^{-1}$ and intercept $-\log(\lambda)$.

Moreover, it is known that if $T \sim W(\alpha, \lambda)$, then $Y = \log(T)$ has an *extreme value* (EV) distribution with location $\mu = -\log(\lambda)$ and scale $\sigma = \alpha^{-1}$ and pdf given by (for $y \in \mathbb{R}$)

$$f_Y(y) = \sigma^{-1} \exp\left[\frac{y - \mu}{\sigma} - \exp\left(\frac{y - \mu}{\sigma} \right) \right],$$

where $\sigma > 0$ and $\mu \in \mathbb{R}$. Thus, Y can be represented by

$$Y = \mu + \sigma Z,$$

where Z is a standard ($\mu = 0$ and $\sigma = 1$) EV random variable. Applying this case in (1.6), one obtain [10]

$$h(t|\boldsymbol{x}) = h_0(t) \exp(\boldsymbol{x}^\top \boldsymbol{\beta}) = \alpha \lambda^\alpha t^{\alpha-1} [(e^{\boldsymbol{x}^\top \boldsymbol{\beta}})^{\frac{1}{\alpha}}]^\alpha = \alpha \underbrace{[\lambda (e^{\boldsymbol{x}^\top \boldsymbol{\beta}})^{\frac{1}{\alpha}}]^\alpha}_{\tilde{\lambda}} t^{\alpha-1}$$

$$= \alpha \tilde{\lambda}^\alpha t^{\alpha-1}.$$

Therefore, if $T \sim W(\alpha, \tilde{\lambda})$, then $Y = \tilde{\mu} + \sigma Z$, where $\sigma = \alpha^{-1}$,

$$\tilde{\mu} = -\log(\tilde{\lambda}) = \underbrace{-\log(\lambda)}_{\beta_0^*} + \underbrace{(-1/\alpha)\, \boldsymbol{x}^\top \boldsymbol{\beta}}_{\boldsymbol{x}^\top \boldsymbol{\beta}^*}$$

$$= \beta_0^* + \boldsymbol{x}^\top \boldsymbol{\beta}^*,$$

$\beta_0^* = -\log(\lambda)$ and $\boldsymbol{\beta}^* = -\sigma \boldsymbol{\beta}$. Finally,

$$Y = \beta_0^* + \boldsymbol{x}^\top \boldsymbol{\beta}^* + \sigma Z.$$

1.3.2.   Accelerated Failure Time Model

This model is characterized by taking $Y = \log(T)$, where T is a random variable for describing the failure time, by means of the linear relation

$$Y = \boldsymbol{x}^\top \boldsymbol{\beta}^* + Z^*,$$

where Z^* has a known distribution. This equation is also called the *log-linear regression model*. Thus,

$$T = \exp(Y) = \exp(\boldsymbol{x}^\top \boldsymbol{\beta}^*) \underbrace{\exp(Z^*)}_{T^*} = \exp(\boldsymbol{x}^\top \boldsymbol{\beta}^*)\, T^*.$$

In this case, let $h_0^*(t^*)$ be the hrf of T^*. According to Tableman and Kim (2005) [10], the hrf of T given a covariate vector $\boldsymbol{x}$ can be expressed in terms of $h_0^*(\cdot)$,

$$h(t|\boldsymbol{x}) = h_0^*(\exp(-\boldsymbol{x}^\top \boldsymbol{\beta}^*)\, t)\, \exp(-\boldsymbol{x}^\top \boldsymbol{\beta}^*).$$

Note that the regression model previously discussed can be understood as a special case of this approach.

1.4.   SPECIAL FUNCTIONS

We will use throughout the book several special functions, discussed in the remainder of this section. For more details, we recommend the books written by Gradshteyn and Ryzhik [11] and Abramowitz and Stegun [12] as well as the excellent lecture notes (Engineering Course ME755 - Special Functions. University of Waterloo) of Prof. J. R. Culham, `http://www.mhtlab.uwaterloo.ca/courses/me755/`.

(a) **The gamma function and its incomplete versions:**

The (complete) gamma function is defined as follows: Let $z \in \mathbb{R} - \mathbb{Z}_-$,

$$\Gamma(z) = \int_0^\infty t^{z-1}\, \mathrm{e}^{-t}\, \mathrm{d}t.$$

Setting $t = y^2$ and $\mathrm{e}^{-t} = y$, we have

$$\Gamma(z) = 2 \int_0^\infty y^{2z-1}\, \mathrm{e}^{-y^2}\, \mathrm{d}y \quad \text{and} \quad \Gamma(z) = \int_0^1 (-\log y)^{z-1}\, \mathrm{d}y,$$

respectively. An important recurrence formula is $\Gamma(z+1) = z\,\Gamma(z)$.

The polygamma function is defined by

$$\psi^{(n)}(x) = \frac{\mathrm{d}^{n+1} \log \Gamma(x)}{\mathrm{d}x^{n+1}}.$$

In particular, the digamma function is given by

$$\psi(x) = \psi^{(0)}(x) = \frac{\mathrm{d} \log \Gamma(x)}{\mathrm{d}x}.$$

This function satisfies the relation $\psi(x+1) = x^{-1} + \psi(x)$. Let $\gamma = \lim_{n\to\infty} \sum_{k=1}^n \frac{1}{k} - \log(n) = 0.5772156\ldots$ be the *Euler-Mascheroni constant*. The following integral representations hold

$$\psi(z) = -\gamma + \int_0^1 \frac{1 - t^{z-1}}{1 - t}\, \mathrm{d}t$$

and

$$\psi(z) = -\gamma + \int_0^\infty \frac{(1+t)^{-1} - (1+t)^{-z}}{t}\, \mathrm{d}t.$$

The gamma function is extended by the incomplete gamma function, $\gamma(z, x)$, and its counterpart, $\Gamma(z, x)$, defined by

$$\gamma(z, x) = \int_0^x t^{z-1}\, e^{-t}\, dt \text{ and } \Gamma(z, x) = \int_x^\infty t^{z-1}\, e^{-t}\, dt, \qquad (1.7)$$

respectively, so that $\gamma(z, x) + \Gamma(z, x) = \Gamma(z)$. The functions $\gamma(\cdot, \cdot)$ and $\Gamma(\cdot, \cdot)$ are used as $\gamma_1(z, \cdot) = \Gamma(z)^{-1}\gamma(z, \cdot)$ and $\Gamma_1(z, \cdot) = \Gamma(z)^{-1}\Gamma(z, \cdot)$, so that $\gamma_1(z, x) + \Gamma_1(z, x) = 1$. The following properties hold:

(i) $\gamma(z + 1, x) = z\,\gamma(z, x) - x^z\, e^{-x}$;

(ii) $\Gamma(z + 1, x) = z\,\Gamma(z, x) + x^z\, e^{-x}$;

(iii) $\frac{\Gamma(z+n,x)}{\Gamma(z+n)} = \frac{\Gamma(z,x)}{\Gamma(z)} + e^{-x} \sum_{k=0}^{n-1} \frac{x^{z+k}}{\Gamma(z+k+1)}$;

(iv) $\frac{d\gamma(z,x)}{dx} = -\frac{d\Gamma(z,x)}{dx} = x^{z-1}\, e^{-x}$.

Moreover, the following relations link the gamma incomplete function to the hypergeometric confluent function (which will be defined in the item (e))

$$\gamma(s, z) = s^{-1} z^s\, e^{-z}\, M(1, s + 1, z) = s^{-1} z^s\, M(s, s + 1, -z)$$

and

$$\Gamma(s, z) = e^{-z}\, U(1 - s, 1 - s, z) = e^{-z} z^s\, U(1, 1 + s, z),$$

where $M(\cdot, \cdot, \cdot)$ is the Kummer's first kind confluent hypergeometric function,

$$M(1, s+1, z) = 1 + \frac{z}{(s+1)} + \frac{z^2}{(s+1)(s+2)} + \frac{z^3}{(s+1)(s+2)(s+3)} + \cdots,$$

$U(\cdot, \cdot, \cdot)$ is the Kummer's second kind confluent hypergeometric function and

$$U(1 - s, 1 - s, z) = \frac{z^s}{\Gamma(1 - s)} \int_0^\infty \frac{e^{-u}}{u^s\,(z + u)}\, du.$$

(b) **The error function and its counterpart:**

The error function and its counterpart can be important to solve problems in several areas, such as in heat conduction, probability theory and financial mathematics (as part of Black-Scholes formula). An important property of the error function is its relation with the *normal* distribution defined as follows: If X follows the normal model, say $X \sim N(\mu, \sigma^2)$, its pdf and cdf are given by

$$f_X(x) = (2\pi\sigma^2)^{-1/2} \exp\left[-\frac{1}{2}\left(\frac{x - \mu}{\sigma}\right)^2\right] = \sigma^{-1} \phi\left(\frac{x - \mu}{\sigma}\right)$$

and

$$\Phi\left(\frac{x-\mu}{\sigma}\right) = \int_{-\infty}^{x} f_X(t)\, dt,$$

where $\phi(x) = (2\pi)^{-1}\, e^{-x^2/2}$ and $\Phi(x) = \int_{-\infty}^{x} \phi(t)\, dt$.

The error function is defined as $\mathrm{erf}(x) = 2\,\Phi(\sqrt{2}x) - 1$ or, more clearly, $\mathrm{erf}(x) : (0,\infty) \to [0,1]$

$$\mathrm{erf}(x) = \frac{2}{\sqrt{\pi}} \int_0^x e^{-t^2} dt.$$

Moreover, its counterpart is given by

$$\mathrm{erfc}(x) = 1 - \mathrm{erf}(x) = \frac{2}{\sqrt{\pi}} \int_x^{\infty} e^{-t^2} dt.$$

(c) **The beta function and associated results:**

An important definite integral associated with the gamma function is the beta function

$$B(a,b) = \int_0^1 t^{a-1} (1-t)^{b-1}\, dt, \qquad\qquad (1.8)$$

for $a, b > 0$. From minor algebraic manipulations,

$$B(a,b) = \frac{\Gamma(a)\Gamma(b)}{\Gamma(a+b)}, \quad a, b > 0.$$

Another important function is the incomplete beta function

$$B_x(a,b) = \int_0^x t^{a-1} (1-t)^{b-1}\, dt.$$

We can define the cdf of the beta model, $I_x(a,b) = B_x(a,b)/B(a,b)$, which gives the symmetry relation $I_x(a,b) = 1 - I_{1-x}(b,a)$ and the relation with the hypergeometric function $I_x(a,b) = a^{-1}\, x^a\, F(a, 1-b; a+1; x)$, see item (e).

(d) **The Riemann's zeta function:**

Let x be a complex number. We define the Riemann's zeta function, $\eta(x)$, by

$$\eta(x) = \sum_{n=1}^{\infty} \frac{1}{n^x},$$

where $x = \sigma + it$ with $\sigma > 1$. Notice $|n^x| = n^\sigma$ and, therefore, the series converges for $\sigma > 1$.

The function $\eta(x)$ can also be defined by an infinite product called as *Euler Product*:

$$\eta(x) = \prod_{n=1}^{\infty} \left(1 - \frac{1}{n^x}\right).$$

(e) **The confluent hypergeometric function:**

The hypergeometric function–denoted by $F(a, b; c; x)$–is defined by

$$F(a, b; c; x) = {}_2F_1(a, b; c; x) = \sum_{n=0}^{\infty} \frac{(a)_n (b)_n}{(c)_n} \frac{x^n}{n}$$

$$= \frac{\Gamma(c)}{\Gamma(a)\Gamma(b)} \sum_{n=0}^{\infty} \frac{\Gamma(a+n)\Gamma(b+n)}{\Gamma(c+n)} \frac{x^n}{n!},$$

where $|x| < 1$, $c \neq 0, -1, -2, \ldots$ and $(z)_n$ is the Pochhammer polynomial given by

$$(z)_n = \begin{cases} z(z+1)(z+2)\cdots(z+n-1) = \frac{\Gamma(z+n)}{\Gamma(z)}, & n > 0 \\ 1, & n = 0. \end{cases}$$

This function can be a solution of the hypergeometric differential equation

$$x(1-x)\,y'' + [c - (a+b+1)x]\,y' - a\,b\,y = 0.$$

Two important properties of $F(\cdot, \cdot; \cdot; \cdot)$ are

$$\frac{\mathrm{d}^k F(a, b; c; x)}{\mathrm{d}x^k} = \frac{(a)_k (b)_k}{(c)_k} F(a+k, b+k; c+k; x),$$

for $k = 1, 2, 3, \ldots$, and the integral representation

$$F(a, b; c; x) = \frac{\Gamma(c)}{\Gamma(b)\Gamma(c-b)} \int_0^1 t^{b-1} (1-t)^{c-b-1} (1-xt)^{-a}\mathrm{d}t,$$

for $|x| \leq 1$.

An important special case of $F(a, b; c; x)$ is the *confluent hypergeometric function* (*Kummer*'s function), denoted by $M(a; c; x)$ (of the first kind) and $U(a; c; x)$ (of the second kind). The function $M(\cdot; \cdot; \cdot)$ is

$$M(a; c; x) = \lim_{b \to \infty} F(a, b; c; x/b) = {}_1F_1(a; c; x) = \sum_{i=0}^{\infty} \frac{(a)_i}{(c)_i} \frac{x^i}{i!}.$$

The function $U(\cdot,\cdot,\cdot)$ is

$$U(a;c;x) = \frac{\pi}{\sin(c\pi)}\left[\frac{M(a;c;x)}{\Gamma(1+a-c)\Gamma(c)} - \frac{x^{1-c}\,M(1+a-c;2-c;x)}{\Gamma(a)\Gamma(2-c)}\right],$$

for $c \neq 0, -1, -2, \ldots.$ Two useful integral representations for the first and second kind Kummer functions are

$$M(a;c;x) = \frac{\Gamma(c)}{\Gamma(a)\Gamma(c-a)}\int_0^1 e^{xt}\, t^{a-1}\,(1-t)^{c-a-1}\,\mathrm{d}t,$$

for $c > a > 0$, and

$$U(a;c;x) = \frac{1}{\Gamma(a)}\int_0^\infty e^{xt}\, t^{a-1}(1+t)^{c-a-1}\mathrm{d}t,$$

for $x > 0$ and $c > a > 0$.

(f) **The Meijer G-function:**

The Meijer G-function is defined by

$$G_{p,q}^{m,n}(z) = G_{p,q}^{m,n}\left(z\,\middle|\,\begin{matrix}\boldsymbol{a_p}\\\boldsymbol{b_q}\end{matrix}\right) = G_{p,q}^{m,n}\left(z\,\middle|\,\begin{matrix}a_1,\ldots,a_p\\b_1,\ldots,b_q\end{matrix}\right)$$

$$= \frac{1}{2\pi i}\int_L \frac{\displaystyle\prod_{j=1}^{m}\Gamma(b_j+t)\prod_{j=1}^{n}\Gamma(1-a_j-t)}{\displaystyle\prod_{j=n+1}^{p}\Gamma(a_j+t)\prod_{j=m+1}^{p}\Gamma(1-b_j-t)}\, z^{-t}\mathrm{d}t, \quad (1.9)$$

where $i = \sqrt{-1}$ is the complex unit and L is a suitable closed contour in the complex plane (or the Riemann sphere) (see Section 9.3 in Gradshteyn and Ryzhik (2000) [11] for a description of this path). Special functions and many products of special functions either are G-functions or are expressible as products of G-functions with elementary functions. For example, (1) the exponential function $e^x = G_{0,1}^{1,0}\left(-x\,\middle|\,1\right)$, (2) the Bessel function $J_v(2x) = x^{-v}\,G_{0,2}^{1,0}\left(x^2\,\middle|\,2\ \ 0\right)$, and (3) products of Bessel functions $J_{2\tau}(x) \times J_{2v}(x) = \pi^{-1/2}\,G_{2,4}^{1,2}\left(x^2\,\middle|\,\begin{matrix}\frac{1}{2}\ \ 0\\\tau+v\ \ v-\tau\ \ -\tau-v\end{matrix}\right).$

(g) **The Lauricella function of type A:**

The Lauricella function of type A (Exton, 1978 [13]) is defined by

$$
F_A^{(n)}\left(a; b_1, \ldots, b_n; c_1, \ldots, c_n; x_1, \ldots, x_n\right)
$$
$$
= \sum_{m_1=0}^{\infty} \cdots \sum_{m_n=0}^{\infty} \frac{(a)_{m_1+\cdots+m_n} (b_1)_{m_1} \cdots (b_n)_{m_n}}{(c_1)_{m_1} \cdots (c_n)_{m_n}} \frac{x_1^{m_1} \cdots x_n^{m_n}}{m_1! \cdots m_n!}. \quad (1.10)
$$

(h) **The generalized Kampé de Fériet:**

The generalized Kampé de Fériet function (Exton, 1978 [13]) is defined by

$$
F_{C:D}^{A:B}\left((a):(b_1);\ldots,(b_n);(c):(d_1);\ldots,(d_n);x_1,\ldots,x_n\right)
$$
$$
= \sum_{m_1=0}^{\infty} \cdots \sum_{m_n=0}^{\infty} \frac{((a))_{m_1+\cdots+m_n} ((b_1))_{m_1} \cdots ((b_n))_{m_n}}{((c))_{m_1+\cdots+m_n} ((d_1))_{m_1} \cdots ((d_n))_{m_n}} \frac{x_1^{m_1} \cdots x_n^{m_n}}{m_1! \cdots m_n!},
$$
$$
(1.11)
$$

where $a = (a_1, a_2, \ldots, a_A)$, $b_i = (b_{i,1}, b_{i,2}, \ldots, b_{i,B})$ for $i = 1, 2, \ldots, n$, $c = (c_1, c_2, \ldots, c_C)$, $d_i = (d_{i,1}, d_{i,2}, \ldots, d_{i,D})$ for $i = 1, 2, \ldots, n$, and $((f))_k = ((f_1, f_2, \ldots, f_p))_k = (f_1)_k (f_2)_k \cdots (f_p)_k$ denotes the product of ascending factorials.

(i) **The complex parameter Wright generalized hypergeometric:**

The complex parameter Wright generalized hypergeometric function with p numerator and q denominator parameters: This function (Kilbas *et al.*, 2006, Equation (1.9) [14]) denoted by $_p\Psi_q(\cdot)$ is defined by the power series

$$
_p\Psi_q \left[\begin{array}{c} (\alpha_1, A_1), \ldots, (\alpha_p, A_p) \\ (\beta_1, B_1), \ldots, (\beta_q, B_q) \end{array} ; z \right] = \sum_{n=0}^{\infty} \frac{\displaystyle\prod_{j=1}^{p} \Gamma\left(\alpha_j + A_j n\right)}{\displaystyle\prod_{j=1}^{q} \Gamma\left(\beta_j + B_j n\right)} \frac{z^n}{n!} \quad (1.12)
$$

for $z \in \mathbb{C}$, where $\alpha_j, \beta_k \in \mathbb{C}$, $A_j, B_k \neq 0$ and the series converges for $1 + \sum_{j=1}^{q} B_j - \sum_{j=1}^{p} A_j > 0$.

1.5. STATISTICAL FUNCTIONS

(a) **Ordinary and central moments (Skewness and Kurtosis)**

Let X be a random variable on the probability space $(\Omega, \mathcal{F}, P)$. The expected value of X^k with k integer is called *the kth moment about zero* denoted by

$$
\mu_k' = \mathrm{E}(X^k).
$$

The kth moment about $\mu = \mu'_1$ (called *mean of a probability distribution*) is denoted by μ_k (known as the *kth central moment*) and defined by

$$\mu_k = \mathrm{E}[(X - \mu)^k] = \mathrm{E}\{[X - \mathrm{E}(X)]^k\} = \sum_{s=0}^{k} \binom{k}{s}(-1)^s \mu^s \mu'_{k-s}.$$

By this last result, the following identities are obtained

$$\mu_2 = \mu'_2 - \mu^2, \mu_3 = \mu'_3 - 3\mu'_2\mu + 2\mu^3,$$
$$\mu_4 = \mu'_4 - 4\mu'_3\mu + 6\mu'_2\mu^2 - 3\mu^4,$$
$$\mu_5 = \mu'_5 - 5\mu'_4\mu + 10\mu'_3\mu^2 - 10\mu_2\mu^3 + 4\mu^5,$$

etc. In this context, one can note that (1) $\mu_1 = 0$, (2) $\mu_2 = \mathrm{Var}(X) = \mathrm{E}(X - \mu)^2 = \sigma^2$ is the variance of X (on which can be defined other important quantities, such as the *standard deviation* σ and the *coefficient of variation* $\mathrm{CV}(X) = \sigma/\mu$), (3) $\gamma_1 = \mathrm{skewness}(X) = \mathrm{E}\left(\frac{X-\mu}{\sigma}\right)^3 = \mu_3/\sigma^3$ is the skewness measure and (4) $\gamma_2 = \mathrm{kurtosis}(X) = \mathrm{E}\left(\frac{X-\mu}{\sigma}\right)^4 - 3 = \mu_4/\sigma^4 - 3$ is the kurtosis measure.

The pth descending factorial moment of X is

$$\mu'_{(p)} = \mathrm{E}[X^{(p)}] = \mathrm{E}[X(X-1) \times \cdots \times (X-p+1)] = \sum_{k=0}^{p} s(p,k)\,\mu'_k,$$

where $s(r,k) = (k!)^{-1}[d^k x^{(r)}/dx^k]_{x=0}$ is the *Stirling number* of the first kind. So, the factorial moments can be determined from the ordinary moments given previously.

For empirical studies, the shapes of many distributions can be investigated by the incomplete moments. They play an important role for measuring income quantiles and Lorenz and Bonferroni curves, which will be discussed later.

(b) **Cumulant and generating functions**

Let X be a random variable on $(\Omega, \mathcal{F}, P)$ and assume that $\mathrm{E}(e^{tX}) < \infty$ for $|t| < T < \infty$ and $T > 0$, the moment generating function (mgf) of X is defined by

$$M_X(t) = \mathrm{E}(e^{tX}) = \mathrm{E}\left(1 + \sum_{r=1}^{\infty} t^r X^r / r!\right) = 1 + \sum_{r=1}^{\infty} \mu'_r t^r / r!.$$

From the last identity, the following relation between μ'_r and the mgf holds:

$$\mu'_r = \left.\frac{\partial^r M_X(t)}{\partial t^r}\right|_{t=0}.$$

The following properties are obtained from the definition: Let $X_1, \ldots, X_n$ be mutually independent random variables on $(\Omega, \mathcal{F}, P)$,

(1) $M_{X_i+k}(t) = e^{kt} M_{X_i}(t)$,

(2) $M_{X_1+X_2}(t) = M_{X_1}(t) M_{X_2}(t)$, $M_{X_1-X_2}(t) = M_{X_1}(t) M_{X_2}(-t)$ and $M_{X_1+\cdots+X_k}(t) = M_{X_1}(t) \times \cdots \times M_{X_k}(t)$ for $k \leq n$ and

(3) $E[e^{(X-\mu)t}] = e^{-\mu t} M_X(t)$ (which is known as *central moment generating function (cmgf)*).

Consider that $M_X(t)$ is well-defined. Then, the cumulant generating function (cgf) exists and it is given by

$$K_X(t) = \log[M_X(t)] = \sum_{r=1}^{\infty} \kappa_r \frac{t^r}{r!},$$

where κ_r is called the rth cumulant of X. Similarly to $M_X(t)$, the following relation holds

$$\kappa_r = \left.\frac{\partial^r K_X(t)}{\partial t^r}\right|_{t=0}.$$

The cumulants (κ_s) of X can be determined recursively as

$$\kappa_s = \mu'_s - \sum_{k=1}^{s-1} \binom{s-1}{k-1} \kappa_k \, \mu'_{s-k},$$

where $\kappa_1 = \mu'_1$. Thus, $\kappa_2 = \mu'_2 - \mu'^2_1$, $\kappa_3 = \mu'_3 - 3\mu'_2\mu'_1 + 2\mu'^3_1$, $\kappa_4 = \mu'_4 - 4\mu'_3\mu'_1 - 3\mu'^2_2 + 12\mu'_2\mu'^2_1 - 6\mu'^4_1$, etc. The skewness $\gamma_1 = \kappa_3/\kappa_2^{3/2}$ and kurtosis $\gamma_2 = \kappa_4/\kappa_2^2$ can be evaluated from the third and fourth standardized cumulants.

(c) **Characteristic function**

Let $X \sim F_X$ be a random variable. Its characteristic function (cf) is defined as a complex-valued function $\phi_X(t)$ on a real line given by

$$\phi_X(t) = E(e^{itX}) = \int_{\mathcal{X}} e^{itx} \, dF_X(x) = E[\cos(tX)] + i\,E[\sin(tX)]$$

$$= \int_{\mathcal{X}} \cos(tx) \, dF_X(x) + i \int_{\mathcal{X}} \sin(tx) \, dF_X(x),$$

where $\mathcal{X}$ is the support of X. The cf has the following properties: (1) $|\phi_X(t)| \leq \phi_X(0) = 1$, (2) $\phi_X(-t) = \overline{\phi}_X(t)$, (3) $\phi_X(t)$ is uniformly continuous at t, (4) $\phi_{aX+b}(t) = e^{itb}\phi_X(at)$, (5)

$$\phi_{\sum_{i=1}^{n} X_i}(t) = \prod_{i=1}^{n} \phi_{X_i}(t)$$

for $\{X_i; i = 1, \ldots, N\}$ as a sample (iid) obtained from X,

$$(6) \; \frac{\partial^r \phi_X(t)}{\partial t^r} = i^r \, \mathrm{E}(X^r \, e^{itX}) \text{ and } (7) \; \mu'_r = i^{-r} \left. \frac{\partial^r \phi_X(t)}{\partial t^r} \right|_{t=0}.$$

A further representation for the cf of X is

$$\phi(t) = \int_0^\infty \cos(tx) \, f(x)\mathrm{d}x + i \int_0^\infty \sin(tx) \, f(x)\mathrm{d}x.$$

Based on the power series

$$\cos(tx) = \sum_{r=0}^{\infty} \frac{(-1)^r}{(2r)!}(tx)^{2r} \text{ and } \sin(tx) = \sum_{r=0}^{\infty} \frac{(-1)^r}{(2r+1)!}(tx)^{2r+1},$$

we can write

$$\phi(t) = \sum_{r=0}^{\infty} \frac{(-1)^r \, t^{2r}}{(2r)!} \mu'_{2r} + i \sum_{r=0}^{\infty} \frac{(-1)^r \, t^{2r+1}}{(2r+1)!} \mu'_{2r+1}.$$

(d) **Probability Weighted Moment**

Greenwood *et al.* [15] defined the *probability weighted moments* (PWMs). Let $X \sim F_X$ be a strictly positive random variable on $(\Omega, \mathcal{F}, P)$ with support $\mathcal{X}$. The PWMs of X are given by

$$\tau_{l,j,k} = \mathrm{E}[X^l \, F(X)^j \, \overline{F}(X)^k] = \int_{\mathcal{X}} x^l \, F(x)^j \, \overline{F}(x)^k \, \mathrm{d}F(x),$$

where l, j, k are positive integer numbers. As a result, they proved that $\tau_{l,j,k}$ is proportional to the lth moment about zero of the $(j+1)$th order statistics from a random sample of size $k + j + 1$,

$$\tau_{l,j,k} = B(j+1, k+1) \, \mathrm{E}[X^l_{(j+1):(k+j+1)}].$$

By according to recent literature, we set $\tau_{s,r} = \tau_{s,0,r}$ for the (s,r)th PWM of X defined by $\tau_{s,r} = \mathrm{E}[X^s G(X)^r]$ for $s, r = 0, 1, \ldots$.

(e) **L-Moments**

L-moments (Hosking (1990) [16]) are alternatives to moments for measuring the location, scale and shape of probability distributions. The rth *L*-moment of a random variable X with cdf F and quantile function (qf) Q is

$$\lambda_r = \text{E}[X\,P^*_{r-1}\{F(x)\}] = \int_0^1 P^*_{r-1}(u)Q(u)\mathrm{d}u,$$

where $P^*_r(\cdot)$ is the rth shifted Legendre polynomial,

$$P^*_r(u) = \sum_{k=0}^r (-1)^{r-k}\binom{k}{k}\binom{r+k}{k}u^k.$$

In particular, λ_1 is the mean, a location measure, and λ_2 is a scale measure, equals to one half of Gini's mean difference. The dimensionless *L*-moment ratios $\tau_3 = \lambda_3/\lambda_2$ and $\tau_4 = \lambda_4/\lambda_2$ are measures of skewness and kurtosis, respectively. Distributions with maximum entropy subject to constraints on their *L*-moments were proposed by Hosking (2007) [17].

(f) **Mean deviations**

Let X be a random variable having positive support on the probability space $(\Omega, \mathcal{F}, P)$. The amount of scatter in X is measured to some extent by the totality of deviations from the mean and median. Two main related measures are the mean deviations about the mean and median, defined by

$$\delta_1(X) = \int_0^\infty |x - \mu| f(x)\mathrm{d}x \qquad \text{and} \qquad \delta_2(X) = \int_0^\infty |x - M| f(x)\mathrm{d}x,$$

respectively, where $\mu = \text{E}(X)$ and $M = \text{Median}(X)$ denotes the median. The measures $\delta_1(X)$ and $\delta_2(X)$ can be calculated using the following relationships

$$\delta_1(X) = \int_0^\mu (\mu - x)f(x)\mathrm{d}x + \int_\mu^\infty (x - \mu)f(x)\mathrm{d}x$$

$$= 2\,\mu\,F(\mu) - 2\,\mu + 2\int_\mu^\infty x f(x)\mathrm{d}x,$$

and

$$\delta_2(X) = \int_0^M (M - x)f(x)\mathrm{d}x + \int_M^\infty (x - M)f(x)\mathrm{d}x$$

$$= 2\int_M^\infty x f(x)\mathrm{d}x - \mu.$$

(g) **Income functions**

Let X be random variable with cdf F. The Bonferroni curve (BC) of X is defined by

$$B_F(p) = \frac{1}{p\,\mu} \int_0^p F^{-1}(t)\, \mathrm{d}t,$$

where $p \in (0, 1]$ and $F^{-1}(t) = \inf\{x; F(x) \geq t\}$. The quantity $B_F(p)$ is represented in the orthogonal plane $[F(x), B_F(F(x))]$ within a unit square.

Notice that, for $p \to 0$, $B_F(p)$ has the form $0/0$; *i.e.*, BC does not always start from the origin.

Moreover, since $\partial B/\partial p > 0$, $B_F(p)$ is strictly increasing, but–as one can not say about its second derivative–$B_F(p)$ can take both convex and concave forms for the same parametric point.

The Bonferroni index, B, is defined as the area limited by the ordinate axis and the B_F curve, given by

$$B = 1 - \int_0^1 B_F(p)\, \mathrm{d}p.$$

The relation between the B_F curve and mean residual lifetime $\epsilon_F(x)$ (a useful quantity in reliability) is

$$B_F(F(x)) = \frac{1}{F(x)} - \frac{1}{\mu}\frac{\bar{F}(x)}{F(x)}[\epsilon_F(x) + x],$$

where

$$\epsilon_F(x) = \frac{\mu\,[1 - F(x)B_F(F(x))]}{\bar{F}(x)} - x.$$

The Lorenz curve (LC) is defined by

$$L_F(p) = \frac{1}{\mu} \int_0^p F^{-1}(t)\, \mathrm{d}t$$

and the Gini index (GI) by

$$G = 1 - 2 \int_0^1 L_F(p)\, \mathrm{d}p.$$

For a positive random variable X, the Bonferroni and Lorenz curves are defined by $B(p) = m_1(q)/(p\,\mu_1')$ and $L(p) = m_1(q)/\mu_1'$, respectively, where $m_n(q) = \int_0^q x^n f(x)\, \mathrm{d}x$, $q = Q(p)$ is the qf of X in p and p is a given probability.

In economics, if p is the proportion of units whose income is lower than or equal to q, $L(p)$ gives the proportion of total income volume accumulated by the set of units with an income lower than or equal to q. The Lorenz curve is increasing and convex and given the mean income, the density function of X can be obtained from the curvature of $L(p)$. In a similar manner, the Bonferroni curve $B(p)$ gives the ratio between the mean income of this group and the mean income of the population. In summary, $L(p)$ yields fractions of the total income, whereas the values of $B(p)$ refer to relative income levels.

(h) Entropies

The seminal idea about Information Theory was proposed by Hartley (1928) [18], who pioneered a logarithmic measure of information for communication. Subsequently, Shannon (1948) [19] formalized this idea by defining the entropy and mutual information concepts. The relative entropy notion (which would later be called *divergence*) was proposed by Kullback and Leibler (1951) [20]. The Kullback-Leibler's measure can be understood like a comparison criterion between two distributions. In this section, we present a discussion on two classes of entropy measures. Let X be a random variable with cdf and pdf $F_X(x)$ and $f_X(x)$:

(h.1) Rényi entropy

The Rényi entropy of X with order β is

$$
\begin{aligned}
H_{\mathrm{R}}^{\beta}(X) &= \frac{1}{1-\beta} \log\left\{ \mathrm{E}_X\left[f_X(X)^{\beta-1} \right] \right. \\
&= \frac{1}{1-\beta} \log\left[\int_{\mathcal{X}} f_X(x)^{\beta-1} \, \mathrm{d}F_X(x) \right] \\
&= \frac{1}{1-\beta} \times \left\{ \begin{array}{l} \log\left(\int_{\mathcal{X}} f_X(x)^{\beta} \, \mathrm{d}x \right), \ X \text{ is continuous,} \\ \log\left(\sum_{x \in \mathcal{X}} f_X(x)^{\beta} \right), \ X \text{ is discrete,} \end{array} \right.
\end{aligned}
$$

where $\beta \in (0, \infty) \setminus \{1\}$.

(h.2) Shannon entropy

The Shannon entropy of X is

$$
\begin{aligned}
H_{\mathrm{S}}(X) &= \mathrm{E}_X\left\{ -\log[f_X(X)] \right\} = -\int_{\mathcal{X}} \log[f_X(x)] \, \mathrm{d}F_X(x) \\
&= \left\{ \begin{array}{l} -\int_{\mathcal{X}} \log[f_X(x)] \, f_X(x) \, \mathrm{d}x, \ X \text{ is continuous,} \\ -\sum_{x \in \mathcal{X}} \log[f_X(x)] \, f_X(x), \ X \text{ is discrete.} \end{array} \right.
\end{aligned}
$$

Rényi and Shannon entropies for the beta modified Weibull distribution were obtained by Nadarajah *et al.* (2011) [21].

(i) Reliability

This point puts its focus on a quantity $R = P(Y < X)$, which is called *reliability*, where the random variables X and Y are independently distributed. This quantity is very useful in survival analysis; for instance, in the system mechanical reliability context, consider that X represents the strength of a component, which is subject to a stress Y, then R describes the system performance.

A form for estimating R is given as follows: Let $X \sim F_X$ and $Y \sim G_Y$ be two random variables. Assume that the distribution of Y is known and a random sample from X, say $X_1, \ldots, X_n$, is observed. Then,

$$R = P(Y < X) = \int_{-\infty}^{\infty} \int_{-\infty}^{x-} \mathrm{d}G(y)\,\mathrm{d}F(x)$$

$$= \int_{-\infty}^{\infty} G_Y(x-)\,\mathrm{d}F(x) = \mathrm{E}_X\{G_Y(X-)\}.$$

In practice, $G_Y(y)$ is known and, therefore, a plausible procedure is defined by

$$\widehat{R} = \frac{1}{n} \sum_{i=1}^{n} G_Y(X_i-).$$

It may be mentioned here that related problems have been widely used in the statistical literature. The simple expression for calculating the reliability in survival analysis is

$$R = P(X > Y) = \int_{0}^{\infty} f_X(x)\,G_Y(x)\,\mathrm{d}x,$$

where $f_X(x) = \mathrm{d}F_X(x)/\mathrm{d}x$.

(j) Order statistics

Order statistics are commonly used in many fields, such as hydrology, engineering and industry, among others. Further, such quantities follow an important rule as support to Statistical Inference and Nonparametric Statistics. This section tackles on some order statistics results.

Let $X_1, \ldots, X_n$ be a random sample drawn from X having cdf and pdf $F_X(x)$ and $f_X(x)$, respectively. The cdf and pdf of the ith order statistic, say $X_{i:n}$, in a random sample of size n are

$$F_{X_{i:n}}(x) = \sum_{k=i}^{n} \binom{n}{k} F_X(x)^k \left[1 - F_X(x)\right]^{n-k}$$

and

$$f_{X_{i:n}}(x) = \frac{n!}{(i-1)!(n-i)!} f_X(x) F_X(x)^{i-1} \left[1 - F_X(x)\right]^{n-i},$$

respectively. For example, Cordeiro *et al.* (2014) [22] determined the pdf and cdf of the order statistics of the Marshall-Olkin family, among other properties.

(1) **Goodness-of-fit Statistics (GoF)**

Let $\widehat{\boldsymbol{\theta}}$ be the MLE of $\boldsymbol{\theta}$ with dimension k based on an observed sample $x_1, \ldots, x_n$. The Akaike Information Criterion (AIC), Bayesian Information Criterion (BIC) and Consistent Akaike Information Criterion (CAIC) are, respectively, given by

$$\text{AIC} = 2\,k - 2 \sum_{i=1}^{n} \log([f(x_i; \widehat{\boldsymbol{\theta}})])$$

$$\text{BIC} = 2 \log(n) - 2 \sum_{i=1}^{n} \log([f(x_i; \widehat{\boldsymbol{\theta}})])$$

and

$$\text{CAIC} = \text{AIC} + \frac{2\,k\,(k+1)}{n - k - 1}.$$

The GoF measures discussed are based on pdfs, but other quantities have been proposed by means of cdfs, see Chen and Balakrishnan (1995) [23] and Pakyari and Balakrishnan (2012) [24]. As follows, we present a discussion on two of them: The modified Anderson-Darling (A^*) and modified Cramér-von Mises (W^*) statistics. Consider we are interested to test the null hypothesis

$$\mathcal{H}_0 \colon F_X \in \mathcal{F}_\theta \text{ for some } \boldsymbol{\theta} \in \Theta.$$

The A^* and W^* statistics employ the distances

$$D(F_n, F_{\widehat{\boldsymbol{\theta}}}) = n \int_{-\infty}^{\infty} \frac{[F_n(x) - F_{\widehat{\boldsymbol{\theta}}}(x)]^2}{F_{\widehat{\boldsymbol{\theta}}}(x)\,[1 - F_{\widehat{\boldsymbol{\theta}}}(x)]} \, \mathrm{d}F_{\widehat{\boldsymbol{\theta}}}(x)$$

and

$$D(F_n, F_{\widehat{\theta}}) = n \int_{-\infty}^{\infty} [\, F_n(x) - F_{\widehat{\theta}}(x)\,]^2 \, \mathrm{d}F_{\widehat{\theta}}(x),$$

where $F_n(x)$ is the empirical cdf and $F_{\widehat{\theta}}(x)$ is the theoretical cdf under the null hypothesis. We employ W^* and A^* like goodness-of-fit statistics to assess fits in used databases. The measures W^* and A^* were discussed by Chen and Balakrishnan (1995) [23]. As decision rule, the better fits are related with smaller values of these statistics. Let $H(x;\varphi)$ be the cdf such that H is known and φ is unknown. To obtain the statistics W^* and A^*, we can proceed as in Algorithm 1.

Algorithm 1 Sample version for W^* and A^*

1: Compute $v_i = H(x_i; \widehat{\boldsymbol{\theta}})$, where the x_i's are in ascending order and $y_i = \Phi^{-1}(v_i)$, where $\Phi^{-1}(\cdot)$ is the standard normal quantile function;
2: Compute $u_i = \Phi\{(y_i - \bar{y})/s_y\}$, where $\bar{y} = n^{-1}\sum_{i=1}^{n} y_i$ and $s_y^2 = (n-1)^{-1}\sum_{i=1}^{n}(y_i - \bar{y})^2$ and then calculate

$$W^2 = \sum_{i=1}^{n}\left\{ u_i - \frac{(2i-1)}{2n}\right\}^2 + \frac{1}{12n}$$

and

$$A^2 = -n - \frac{1}{n}\sum_{i=1}^{n}\{(2i-1)\log(u_i) + (2n+1-2i)\log(1-u_i)\};$$

3: Use the modified versions $W^* = W^2(1+0.5/n)$ and $A^* = A^2(1+ 0.75/n + 2.25/n^2)$.

Chapter 2

Exponentiated Models

Abstract: The exponentiation transform of cumulative distributions can furnish more flexible models. Such procedure generates the exponentiated G (exp-G) distributions. This chapter presents a survey on the exp-G models and its mathematical properties. In particular, explicit expressions for the quantile function, ordinary and incomplete moments, generating function, income measures, order statistics and entropies are addressed.

Keywords:: Exponentiated model; Generating function; Hazard function; Moment; Weibull distribution.

2.1.   INTRODUCTION

The proposal of more flexible distributions is an activity often required in practical contexts. In particular, adding a positive real parameter (say $\alpha > 0$) to a cdf $G(x)$ by exponentiation gives a cdf $G(x)^\alpha$ that can provide interesting mathematical properties and better fits to data sets in different contexts. Several works have provided evidence that such class covers both monotonic and non-monotonic hazard rates [25, 26]. Despite simplicity of the approached transformation, the resulting distribution is richer than the baseline $G(x)$. Thus, a tailored treatment for this transformation is required.

Let $G(x)$ and $g(x)$ be the cdf and pdf, respectively, of a known random variable (say, a *baseline model*). A random variable X is said to have the *exponentiated-G* ("exp-G" for short) class if its cdf and pdf are

$$F(x) = G(x)^\alpha, \quad \text{for } x \in \mathcal{D} \subset \mathbb{R} \tag{2.1}$$

and

$$f(x) = \alpha \, g(x) \, G(x)^{\alpha-1}, \tag{2.2}$$

Gauss M. Cordeiro, Rodrigo B. Silva & Abraão D. C. Nascimento

respectively. We omit the parametric elements for simplicity. This model is denoted by $X \sim$ exp-G$(\boldsymbol{\theta})$, where $\boldsymbol{\theta} = (\alpha, \boldsymbol{\delta}^\top)^\top \in \boldsymbol{\Theta} \subset \mathbb{R}^{p+1}$, $\top$ is the transposition operator, α represents the additional parameter, $\boldsymbol{\Theta}$ is the parametric space of the generated exp-G distribution and $\boldsymbol{\delta}$ is the p-dimensional vector of parameters of the baseline distribution. As one of its properties, the exp-G class can be understood as the *proportional reversed hazard rate model*. In summary, the reversed hazard rate function (rhrf) is the probability of observing an outcome within a neighborhood of x, conditional on the outcome being no more than x, and it is defined (for any baseline model) by $\lambda_G(x) = \mathrm{d}\{\log[G(x)]\}/\mathrm{d}x = g(x)/G(x)$ [27, 28]. Thus, the exp-G rhrf is

$$\lambda_F(x) = \frac{\alpha\, g(x)\, G(x)^{\alpha-1}}{G(x)^\alpha} = \alpha\, \lambda_G(x),$$

i.e., the rhrf of the exp-G class is a proportional rhrf. Models which satisfy this characteristic have been sought in the lifetime data analysis literature [29, 30]. An important aspect is that the class determined by (2.1) and (2.2) under $\alpha \in \mathbb{N}$ was pioneered by Lehmann (1953) [31], called initially by *Lehmann alternative type I*. The physical interpretation of the additional parameter of the exp-G class is discussed as follows:

- $F(x) = G(x)$ for $\alpha = 1$;

- For $\alpha = n \in \mathbb{N}$, $F(x)$ represents the cdf of the maximum value defined on a n-variate random sample from $Y \sim G$, say $\{Y_1, \ldots, Y_n\}$:

$$X(n) = \max\{Y_1, \ldots, Y_n\}.$$

The subsequent discussion emphasizes the importance of exp-G distributions. Consider a biological situation, on which corrupted cells are battling to provide observable tumours. Set X_j, $j = 1, \ldots, N$, as the time for the jth corrupted cell to become in a observable tumour (promotion time), where N means the latent number of corrupted cells which may furnish the interest event. Admit N having probability mass function (pmf) given by $p_n = \Pr(N = n)$ for $n = 0, 1, \ldots$ Let $A_N(s) = \sum_{n=0}^{\infty} p_n s^n$ be the corresponding probability generating function (pgf) for $0 < s < 1$, and p_0 the cure rate. Conditional on N, assume that the X_j's are independent random variables having a common baseline pdf $g(x)$ and survival function $S(x) = 1 - G(x)$. Given $N = n$ and the lifetime $T = t$, let Z_j, $j = 1, \ldots, n$, be independent random variables, independently of N, following a Bernoulli distribution with success probability $G(t)$ indicating the presence of the jth competing cause at time t. Further, we consider that the population is divided into two sub-populations of cured and non-cured patients defined by: $U_t = 1$ if $Z_1 + \cdots + Z_N = 1$ and $U_t = 0$ if $Z_1 + \cdots + Z_N = 0$, where $\Pr(U_t = 1) = 1 - p_0$. Let T be a non-negative lifetime random variable and X the promotion time with pdf $g(x)$. Define the distribution of T as the conditional distribution of X, given $U_t = 1$. Under this set up, Rodrigues *et al.* (2011) [32] demonstrated that the pdf of T is given by

$$f_T(t) = \frac{g(t)}{1 - p_0} \left\{ \left. \frac{\mathrm{d}A_N(s)}{\mathrm{d}s} \right|_{s=S(t)} \right\}.$$

The corresponding hrf is

$$h_T(t) = \frac{g(t)}{A_N\left(S(t)\right) - p_0} \left\{ \left. \frac{\mathrm{d}A_N(s)}{\mathrm{d}s} \right|_{s=S(t)} \right\}.$$

The class of distributions specified by the pdf $f_T(t)$ is quite broad. It contains as special cases either exp-G distributions or generalizations from them. For example, the generalized exponential Poisson (Barreto-Souza and Cribari-Neto, 2009 [33]), Lehmann alternatives, Weibull-geometric (Barreto-Souza *et al.*, 2010 [34]), exponentiated Weibull (EW) [25], generalized modified Weibull (Carrasco *et al.*, 2008a [5]) and exponential power series (Chahkandi and Ganjali, 2009 [35]) distributions. The properties of the exponentiated distributions were widely discussed in the last years, see Mudholkar and Srivastava (1993) [7] and Mudholkar *et al.* (1995) [36] for exponentiated Weibull, Gupta *et al.* (1998) [37] and Gupta and Kundu (2001) [38] for exponentiated exponential, Nadarajah and Gupta (2007) [39] for exponentiated gamma, Carrasco *et al.* (2008) [5] for exponentiated modified Weibull and Cordeiro *et al.* (2011) [40] for exponentiated generalized gamma distribution.

2.2.   SPECIAL CASES

In this section, we discuss some special cases of the exp-G class. As the first case, Gupta *et al.* [37] pioneered the *exponentiated exponential* (EE) distribution as a generalization of the standard exponential distribution. Its two parameters represent the shape and the scale parameters as those cases of the classical gamma and Weibull distributions. The mathematical properties of the EE distribution were studied by Nadarajah and Kotz [41] and also by Gupta and Kundu [42]. Several papers have addressed other properties: see Gupta and Kundu (2001a, 2001b, 2007) [38, 43, 44], Raqab and Ahsanullah (2001) [45], Raqab (2002) [46], Sarhan (2007) [47], Abdel-Hamid and Al-Hussaini (2009) [48], and Aslam *et al.* (2010) [49]. Four special cases of the exponentiated distributions are discussed in Nadarajah and Kotz (2006) [41]:

the exponentiated gamma (EG) distribution [50], EW [26], exponentiated Gumbel (EGu) [51] and exponentiated Fréchet (EFr) [41]. For statistical inference and regression models, many papers have advanced under the assumption of exponentiated type distributions:

☞ Inferential methods:

- (Gupta and Kundu, 2001 [43], 2007 [44]): Point estimation by means of maximum likelihood, moment, percentile, least squares, Bayesian and L-moment methods. Inference on stress-strength parameter.
- (Gupta and Kundu, 2001 [52]): Bayesian inference.
- (Gupta and Kundu, 1999 [42]): Results in terms of behaviour of the hazard functions of three distributions.
- (Kundu *et al.*, 2005 [53]): Discussion about discrimination between the log-normal and generalized exponential distributions in terms of the Kolmogorov-Smirnov (KS) distance.

☞ Regression model on the exp-G distribution:

- (Hashimoto, 2010 [54]): the log-exponentiated Weibull regression model.

2.2.1. The EE Distribution

In the first half of the nineteenth century, some cdf's were used by Gompertz (1825) [55] and Verhulst (1847) [56] in order to contrast known human mortality tables and the mortality advance. One of them has been determined as follows:

$$F^*(x) = \left(1 - \rho\,e^{-\lambda x} \right)^\alpha, \quad \text{for } x > \frac{1}{\lambda}\log(\rho), \tag{2.3}$$

where ρ, λ and α are all positive real numbers. The generalized exponential (or the EE distribution) is defined as a special case of the Gompertz-Verhulst distribution function (2.3) when $\rho = 1$.

Thus, a random variable X is said to have the EE distribution if its cdf and pdf are

$$F(x) = \left(1 - e^{-\lambda x} \right)^\alpha$$

and

$$f(x) = \alpha\,\lambda\,e^{-\lambda x}\left(1 - e^{-\lambda x} \right)^{\alpha-1}, \tag{2.4}$$

respectively, for $x > 0$ and $\lambda, \alpha > 0$. We write $X \sim \mathrm{EE}(\alpha, \lambda)$. Note that the EE model can be understood as an element from the exp-G class by inserting the exponential cdf and pdf,

$$G(x) = 1 - e^{-\lambda x} \quad \text{and} \quad g(x) = \lambda \, e^{-\lambda x},$$

in equations (2.1) and (2.2). The last redefinition is richer because all the physical formation semantic can be embedded to the applications involving the EE model.

It follows some important properties in closed-form for the EE distribution. Let $X \sim \mathrm{EE}(\alpha, \lambda)$ and $\varphi(x) \colon \mathbb{R} \to \mathbb{R}$ be a Borel measurable function. Then,

$$\mathrm{E}[\varphi(X)] = \alpha \lambda \int_0^\infty \varphi(x) \, e^{-\lambda x} \, (1 - e^{-\lambda x})^{\alpha-1} \, \mathrm{d}x.$$

As special cases, the EE mgf $(M(t))$ and cf $(\phi(t))$ are

$$M(t) = \mathrm{E}(e^{t X}) = \alpha \lambda \int_0^\infty e^{(t-\lambda)\,x} \left[1 - e^{-\lambda x} \right]^{\alpha-1} \mathrm{d}x$$

$$= \alpha \int_0^1 y^{-t/\lambda} \, (1 - y)^{\alpha-1} \mathrm{d}y = \alpha \, B\left(1 - \frac{t}{\lambda}, \alpha \right), \qquad (2.5)$$

and

$$\phi(t) = \mathrm{E}(e^{it X}) = \alpha \, B\left(1 - \frac{i\,t}{\lambda}, \alpha \right),$$

respectively.

According to Nadarajah (2011) [25], also from (2.5), the nth moment of X is

$$\mathrm{E}(X^n) = \frac{(-1)^n \, \alpha}{\lambda^n} \left. \frac{\partial^n}{\partial p^n} B(\alpha, p + 1 - \alpha) \right|_{p=\alpha}. \qquad (2.6)$$

From (2.6), one can derive the ordinary moment, $\mu = \mathrm{E}(X)$, the variance, $\sigma^2 = \mathrm{var}(X) = \mathrm{E}(X - \mu)^2$, the skewness, $\gamma_1 = \mathrm{E}[(\frac{X-\mu}{\sigma})^3]$, and the kurtosis, $\gamma_2 = \mathrm{E}[(\frac{X-\mu}{\sigma})^4]$:

$$\mu = \lambda^{-1} [\psi(\alpha + 1) + C],$$

where C is Euler's constant,

$$\sigma^2 = (6 \, \lambda^2)^{-1} [\pi^2 - 6 \, \psi'(\alpha + 1)],$$

$$\gamma_1 = [\pi^2 - 6 \, \psi'(\alpha + 1)]^{-3/2} \{6 \, \sqrt{6} \, [2 \, \eta(3) + \psi''(\alpha + 1)]\}$$

and

$$\gamma_2 = \frac{9}{5} \frac{3 \, \pi^4 + 60 \, \{\psi'(\alpha + 1)\}^2 - 20 \, \pi^2 \, \psi'(\alpha + 1) - 20 \, \psi'''(\alpha + 1)}{[\pi^2 - 6 \, \psi'(\alpha + 1)]^2}.$$

Next, we give two generalizations of the EE distribution:

The **generalized exponential (GE)** distribution has cdf and pdf derived by Gupta and Kundu (1999) [42] as

$$F(x) = \left[1 - \exp\left(\frac{x - \mu}{\lambda} \right) \right]^{\alpha}$$

and

$$f(x) = \frac{\alpha}{\lambda} \left[1 - \exp\left(\frac{x - \mu}{\lambda} \right) \right]^{\alpha-1} \exp\left(\frac{x - \mu}{\lambda} \right),$$

respectively, where $x > \mu$, $\alpha, \lambda > 0$.

The **beta generalized exponential** distribution pioneered by Barreto-Souza *et al.* (2008) [33] has four parameters and pdf

$$f(x) = \frac{\alpha \lambda}{B(a,b)} \exp(-\lambda x) \left[1 - \exp(-\lambda x) \right]^{\alpha a - 1}$$
$$\times \left\{ 1 - \left[1 - \exp(-\lambda x) \right]^{\alpha} \right\}^{b-1}$$

for $x > 0$, $a > 0$, $b > 0$, $\alpha > 0$ and $\lambda > 0$. It contains, as special cases, the EW, EE and beta exponential (Nadarajah and Kotz, 2006 [41]) distributions, among others. It allows for bathtub shaped, monotonically increasing, monotonically decreasing and upside-down bathtub hazard rates.

2.2.2.   The EW Distribution

The two-parameter Weibull is one of the most known lifetime distributions (Murthy *et al.*, 2004, Rinne, 2009) [57, 58]. Its main drawback is its difficulty to cover non-monotonic hazard rates (e.g., bathtub shaped hazard rates), motivating the proposal of extended models from it. The EW distribution introduced by Mudholkar and Srivastava (1993) [7] and Mudholkar *et al.* (1995) [36] is the first W extension to satisfy non-monotonic hazard rates, particularly bathtub shapes. Evidences have illustrated that the EW distribution may outperform traditional models, such as the exponential, gamma, Weibull and log-normal distributions. Various structural properties and applications of the EW distribution have been studied by many authors, among them, Mudholkar *et al.* (1995) [36], Mudholkar and Hutson (1996) [59], Nassar and Eissa (2003) [60], Pal *et al.* (2006) [61], Nadarajah and Gupta (2005) [62], and

Nadarajah and Kotz (2006) [41]. A random variable X is said to have the EW distribution if its cdf and pdf are

$$F(x) = \left[1 - e^{-(\lambda x)^c}\right]^\alpha \tag{2.7}$$

and

$$f(x) = c\,\alpha\,\lambda^c\,x^{c-1}\,e^{-(\lambda x)^c}\left[1 - e^{-(\lambda x)^c}\right]^{\alpha-1}, \tag{2.8}$$

respectively, for $x > 0$, $c > 0$, $\lambda > 0$ and $\alpha > 0$. We write $X \sim \mathrm{EW}(c, \alpha, \lambda)$. The particular case $c = 1$ is the EE distribution due to Gupta and Kundu (1999) [42]. The case $\alpha = 1$ gives the Weibull distribution. The case $c = 2$ corresponds to the Burr type X distribution studied by Sartawi and Abu-Salih (1991) [63], Kundu and Gupta (2004) [64], Kundu and Raqab (2005) [65], Malinowska and Szynal (2005) [66], Zhou *et al.* (2008) [67], among others. The special case $c = 2$ and $\alpha = 1$ is the Rayleigh distribution. Similarly to the previous section, we now present some properties of the EW model. Nassar and Eissa (2003) [60] derived expressions for the mode of the EW pdf. They provided different expressions depending on the values of c and α. For $\alpha = 1$, then

$$\mathrm{Mode} = \frac{1}{\lambda}\left(\frac{c-1}{c}\right)^{1/c}.$$

If $c < 1$ and $\alpha < 1$, then there is no mode. If $c < 1$, $\alpha > 1$ and $c\alpha > 1$ then

$$\mathrm{Mode} = \frac{1}{\lambda}\left[\frac{2\,(c\,\alpha - 1)}{c(\alpha + 1)}\right]^{1/c}.$$

Finally, if $c < 1$, $\alpha > 1$ and $c\alpha \leq 1$, then the mode is at zero. The qf of the EW distribution is

$$F^{-1}(p) = \frac{1}{\lambda}\left[-\log\left(1 - p^{1/\alpha}\right)\right]^{1/c} \tag{2.9}$$

for $0 < p < 1$. In particular, the median is

$$\mathrm{Median} = \frac{1}{\lambda}\left[-\log\left(1 - 2^{-1/\alpha}\right)\right]^{1/c}.$$

As a consequence, if $U \sim U(0,1)$, then $X = \lambda^{-1}\left[-\log(1 - U^{1/\alpha})\right]^{1/c}$ is a random number from the EW distribution. For $c = 1$, we obtain the EE distribution. The hrf and mrlf of the EW distribution are

$$h(x) = c\,\alpha\,\lambda^c\,x^{c-1}\,\exp\left[-(\lambda x)^c\right]\left\{1 - \exp\left[-(\lambda x)^c\right]\right\}^{-1}$$

and

$$\mu(x) = \frac{\int_x^\infty [1 - \{1 - \exp[-(\lambda y)^c]\}^\alpha]\,dy}{1 - \{1 - \exp[-(\lambda x)^c]\}^\alpha},$$

respectively. The hrf allows for constant, monotonically increasing, monotonically decreasing, unimodal and bathtub shaped hazard rates. In particular,

- bathtub shapes with a unique change point occur when $c > 1$ and $c\alpha < 1$;

- unimodal shapes with a unique change point occur when $c < 1$ and $c\alpha > 1$;

- monotonically increasing shapes occur when $c > 1$ and $c\alpha > 1$ and

- monotonically decreasing shapes occur when $c < 1$ and $c\alpha < 1$. The hazard rate is constant when $c = \alpha = 1$.

Nadarajah (2009) [68] compiles a collection of distributions allowing for bathtub hazard rates. Furthermore, $h(x) \sim c\alpha x^{-1}$ as $x \to 0$ and $h(x) \sim c\lambda^c x^{c-1}$ as $x \to \infty$. So, both lower and upper tails of the hrf behave polynomially. The subsequent discussion involves the derivation of the cf $\phi_X(t)$. Nadarajah and Pogány (2011) [69] were the first ones to provide a general expression of $\phi_X(t)$ for a two-parameter Weibull random variable with shape parameter c and scale parameter λ. They proved that (see Section 1.4)

$$\phi_X(t) = {}_1\Psi_0 \left[\begin{matrix} (1, 1/c) \\ - \end{matrix} ; i\,t/\lambda \right]$$

for $c > 1$ and $i = \sqrt{-1}$. Let X be a random variable having the pdf (2.8) Nadarajah *et al.* [26] obtained an expansion for the EW cf

$$\phi_X(t) = \alpha \sum_{j=0}^\infty \binom{\alpha - 1}{j} \frac{(-1)^j}{j+1} {}_1\Psi_0 \left[\begin{matrix} (1, 1/c) \\ - \end{matrix} ; i\,t/(j+1)^{1/c}\lambda \right] \qquad (2.10)$$

for $c > 1$. Hypergeometric functions are included as in-built functions in most popular mathematical software packages, so the special function in (2.10) can be easily evaluated using procedures in the software packages `Maple`, `Matlab` and `Mathematica`. In the following, some EW extensions are discussed.

The **generalized Weibull (GW)** distribution:

The GW distribution is closely related to the EW distribution and was proposed by Mudholkar *et al.* (1996) [59] and Mudholkar and Sarkar (1999) [70]. It has cdf

$$F(x) = 1 - \left[1 - \exp\{-\alpha(\lambda x)^{1/c}\}\right]^{1/\alpha}$$

for $x > 0$, $\lambda > 0$, $c > 0$ and $\alpha > 0$. The hazard function can behave as bathtub shaped, monotonic and constant hazard rates. Beyond the EW properties, it satisfies the property of proportional hazards. The last fact is sought to deal with multiple samples in the repair-reuse type reliability context.

The **first modified EW (1-MEW)** distribution:

Gera (1997) [71] presented a modification of the EW distribution given by the cdf

$$F^*(x) = \exp\left(-sx^r\right) F(x)$$

for $r > 0$ and $s > 0$, where $F(\cdot)$ is given (2.7). In fact, this model extends the EW distribution. Gera (1997) [71] showed by numerical calculations that the mean time between failures corresponding to F^* is lower than that for F, *i.e.*

$$\int_0^\infty [1 - F^*(x)]\, \mathrm{d}x < \int_0^\infty [1 - F(x)]\, \mathrm{d}x$$

for all parameter values.

The **second modified EW (2-MEW)** distribution:

Gupta *et al.* (1998) [37] proposed a class of distributions specified by the cdf

$$F^*(x) = F(x)^\theta$$

for $\theta > 0$.

The EW law is a special case of this class when replacing F by the Weibull cdf. Gupta *et al.* (1998) [37] derived general properties of this class; for instance, (i) the conditions $\theta > 1$ and $F(\cdot)$ having increasing hazard rates imply that $F^*(\cdot)$ has the increasing hrf and (ii) $\theta < 1$ and $F(\cdot)$ with decreasing hrf make that $F^*(\cdot)$ has increasing hrf.

The **third modified EW (3-MEW)** distribution:

Lai *et al.* (2003) [72] defined the *modified Weibull* (MW) distribution by the pdf

$$f(x) = a(b + \lambda x)x^{b-1}\exp(\lambda x)\exp\left[-ax^b\exp(\lambda x)\right]$$

for $x > 0$, $a > 0$, $b > 0$ and $\lambda > 0$.

Despite this distribution does not extend (2.8), its hrf has increasing and bathtub forms. Lai *et al.* (2003) [72] discussed a data set that reveals that this law can outperform the EW distribution.

The **mixture of EW (MixEW)** distributions:

Cancho and Bolfarine (2001) [73] introduced a mixture form of the EW distribution given by the cdf

$$F^*(x) = (1 - p) + p[1 - F(x)]$$

for $0 < p < 1$, where $F(\cdot)$ is an EW cdf. This distribution is used to model the possibility that long-term survivors are presented in the data. Cancho and Bolfarine (2001) [73] considered inference inference procedures for its parameters using maximum likelihood and Markov Chain Monte Carlo (MCMC) simulation.

The **generalized modified Weibull** distribution:

Carrasco *et al.* (2008a) [5] pionnered a four-parameter generalized modified Weibull (GMW) distribution having pdf

$$f(x) = \frac{\alpha\beta x^{\gamma-1}(\gamma + \lambda x)\exp\left[\lambda x - \alpha x^\gamma\exp(\lambda x)\right]}{\{1 - \exp\left[-\alpha x^\gamma(\lambda x)\right]\}^{1-\beta}}$$

for $x > 0$, $\alpha > 0$, $\gamma > 0$, $\lambda \geq 0$ and $\beta > 0$. Its hrf admits increasing, decreasing, bathtub shaped and upside down bathtub shaped hazard forms. It assumes important lifetime special cases: the Weibull, extreme value, EW and MW distributions, the last one due to Lai *et al.* (2003) [72].

The **odd Weibull** distribution:

Jiang *et al.* (2008) [74] introduced the three-parameter *odd Weibull* distribution specified by the cdf

$$F(x) = 1 - \left\{1 + [\exp\{(\lambda x)^\alpha\} - 1]^\beta\right\}^{-1}$$

for $x > 0$, $\alpha > 0$, $\beta > 0$ and $\lambda > 0$. It allows for increasing, decreasing, constant, bathtub shaped and unimodal hazard rates.

Despite it is not an EW extension, Jiang *et al.* (2008) [74] presented evidences that this model may furnish meaningfully fits.

The **power generalized Weibull** distribution:

Nikulin and Haghighi (2006, 2009) [75, 76] introduced the three-parameter *power generalized Weibull* distribution given by the cdf

$$F(x) = 1 - \exp\left\{1 - [1 + (\lambda x)^\nu]^{1/\gamma}\right\}$$

for $x > 0$, $\nu > 0$, $\gamma > 0$ and $\lambda > 0$. Its hrf covers the forms: increasing, decreasing, constant, bathtub shaped, and unimodal. Note that this distribution does not extends the EW distribution. However, an applied study on censoring in cancer evidences (Nikulin and Haghighi, 2006, 2009 [75, 76]) showed that it can offer an efficient description. Nikulin and Haghighi (2006)[75] also derived a chi-square test for validating the adherence of the generalized power Weibull distribution.

The **exponentiated modified Weibull** distribution:

Voda (2009) [77] proposed the three-parameter *exponentiated modified Weibull* distribution with cdf

$$F(x) = \left[1 - \exp\left(-\frac{x^{2k}}{2a^2}\right)\right]^b$$

for $x > 0$, $k > 0$, $a \geq 0$ and $b > 0$. However, Voda (2009) [77] did not appear to realize that this is in fact a cdf of the EW distribution.

The **beta Weibull** distribution:

Wahed *et al.* (2009) [78] introduced the *beta Weibull* (BW) distribution, a four-parameter generalization of (2.8), given by the pdf

$$f(x) = \frac{\gamma}{\beta B(\alpha_1, \alpha_2)} \left\{1 - \exp\left[-\left(\frac{x}{\beta}\right)^\gamma\right]\right\}^{\alpha_1 - 1}$$
$$\times \exp\left[-\alpha_2 \left(\frac{x}{\beta}\right)^\gamma\right] \left(\frac{x}{\beta}\right)^{\gamma - 1}$$

for $x > 0$, $\alpha_1 > 0$, $\alpha_2 > 0$, $\gamma > 0$ and $\beta > 0$. Its hrf provides the forms: increasing, decreasing, bathtub shaped and upside down bathtub shaped. It contains the EW distribution as a special case for $\alpha_2 = 1$. Cordeiro *et al.* (2012) [79] studied the BW distribution and derived the generating function and the information matrix and investigated the maximum likelihood estimation.

However, Wahed *et al.* (2009) [78] did not appear to realize that the BW distribution was first introduced by Famoye *et al.* (2005) [80] and both these authors and Lee *et al.* (2007) [81]. BW moment-based explicit expressions have initially been proposed by Cordeiro *et al.* (2011) [82]. Other BW general quantities - such as mean deviations, Bonferroni and Lorenz measures, reliability, entropies, distribution of order statistics, and the asymptotic distributions of the extreme values - have been derived by Cordeiro *et al.* (2011) [83].

The **Kumaraswamy Weibull** distribution:

Cordeiro *et al.* (2010) [84] defined the four-parameter *Kumaraswamy Weibull* (KwW) distribution by the pdf

$$f(x) = a\,b\,c\,\lambda^c\,x^{c-1}\exp\left\{-(\lambda x)^c\right\}\left[1 - \exp\left\{-(\lambda x)^c\right\}\right]^{a-1}$$
$$\times \left\{1 - \left[1 - \exp\left\{-(\lambda x)^c\right\}\right]^a\right\}^{b-1}$$

for $x > 0$, $a > 0$, $b > 0$, $\lambda > 0$ and $c > 0$. Some of its special cases are: EW, exponentiated Rayleigh, exponentiated exponential, Weibull, Rayleigh and exponential laws. Its hrf can assume the forms: bathtub shaped, monotonically decreasing, monotonically increasing, and upside down bathtub shaped.

The **beta modified Weibull** distribution:

Silva *et al.* (2010) [85] introduced the five-parameter *beta modified Weibull* (BMW) distribution with pdf

$$f(x) = \frac{\alpha x^{\gamma-1}(\gamma + \lambda x)\exp(\lambda x)}{B\,(a,b)}\exp\left[-b\alpha x^\gamma \exp(\lambda x)\right]$$
$$\times \left\{1 - \exp\left[-\alpha^\gamma \exp(\lambda x)\right]\right\}^{a-1}$$

for $x > 0$, $a > 0$, $b > 0$, $\alpha > 0$, $\gamma > 0$ and $\lambda \geq 0$. It contains as special cases seventeen commonly known distributions including the EW distribution. It allows for bathtub shaped, monotonically decreasing, monotonically increasing, and upside down bathtub shaped hazard rates.

The **generalized beta-generated** class of distributions:

Alexander *et al.* (2011) [86] defined a class of *generalized beta-generated* class of distributions by the pdf

$$f^*(x) = \frac{c}{B(a,b)} f(x) F^{ac-1}(x) \left[1 - F^c(x)\right]^{b-1}$$

for shape parameters $a > 0$, $b > 0$ and $c > 0$, where $F(\cdot)$ is a valid cdf and $f(\cdot)$ is its pdf.

The **exponentiated Kumaraswamy** class of distribution:

Cordeiro and de Castro (2011) [87] pionnered a class of *Kumaraswamy generated* class of distributions by the cdf

$$F^*(x) = 1 - \left[1 - F^\alpha(x)\right]^\beta \tag{2.11}$$

for shape parameters $\alpha > 0$ and $\beta > 0$, where $F(\cdot)$ is a valid cdf. General properties of this class have been studied by Nadarajah *et al.* (2012) [88]. Lemonte *et al.* (2011) [89] introduced the class of *exponentiated Kumaraswamy distributions*, thus generalizing (2.11) to

$$F^*(x) = \left\{ 1 - \left[1 - F^\alpha(x)\right]^\beta \right\}^\gamma \tag{2.12}$$

for shape parameters $\alpha > 0$, $\beta > 0$ and $\gamma > 0$, where $F(\cdot)$ is a valid cdf. Both classes, (2.11) and (2.12), contain the EW distribution as a basic exemplar.

The **exponentiated generalized gamma** distribution:

Cordeiro *et al.* (2012) [79] defined the four-parameter *exponentiated generalized gamma* (EGG) distribution by the pdf

$$f(x) = \frac{\lambda\beta}{\alpha\Gamma(k)} \left(\frac{x}{\alpha}\right)^{\beta k - 1} \exp\left[-\left(\frac{x}{\alpha}\right)^\beta\right] \left\{ \gamma_1\left(k, \left(\frac{x}{\alpha}\right)^\beta\right) \right\}^{\lambda - 1}$$

for $x > 0$, $\alpha > 0$, $\beta > 0$, $k > 0$ and $\lambda > 0$. It includes as special cases the EW, exponentiated generalized half-normal, exponentiated gamma and generalized Rayleigh distributions, among others. It allows for bathtub shaped, monotonically decreasing, monotonically increasing, and upside down bathtub shaped hazard rates.

The **beta extended Weibull** family of distributions:

Cordeiro *et al.* (2011c) [90] defined the *beta extended Weibull* family of distributions by the pdf

$$f(x) = \frac{\alpha h(x)}{B(a,b)} \left\{ 1 - \exp\left[-\alpha H(x)\right] \right\}^{a-1} \exp\left[-\alpha b H(x)\right]$$

for $x > 0$, $a > 0$, $b > 0$ and $\alpha > 0$, where $H(x) \geq 0$ is a monotonic increasing function of x and $h(x) = dH(x)/dx$.

The **Kumaraswamy-generalized gamma** distribution:

Pascoa *et al.* (2011) [91] defined the five-parameter *Kumaraswamy-generalized gamma* (KwGG) distribution by the pdf

$$f(x) = \frac{\lambda \psi \tau}{\alpha \Gamma(k)} \left(\frac{x}{\alpha}\right)^{\tau k - 1} \exp\left[-\left(\frac{x}{\alpha}\right)^{\tau}\right] \left\{ \gamma_1 \left(k, \left(\frac{x}{\alpha}\right)^{\tau}\right) \right\}^{\lambda - 1}$$

$$\times \left(1 - \left\{ \gamma_1 \left(k, \left(\frac{x}{\alpha}\right)^{\tau}\right) \right\}^{\lambda} \right)^{\psi - 1}$$

for $x > 0$, $\alpha > 0$, $\lambda > 0$, $k > 0$, $\psi > 0$ and $\tau > 0$. This distribution contains as special cases the exponentiated generalized gamma due to Cordeiro *et al.* (2012) [79], Kumaraswamy-Weibull and Kumaraswamy-gamma due to Cordeiro and de Castro (2011) [87], and several others. It allows for constant, bathtub shaped, monotonically decreasing, monotonically increasing, and upside down bathtub shaped hazard rates.

2.3. ORDINARY MOMENTS

In general form, if X is an exponentiated type random variable,

$$E(X^k) = \alpha \int_0^{\infty} x^k \, g(x) \, G(x)^{\alpha - 1} \, dx.$$

Setting $u = G(x)$, one has

$$E(X^k) = \alpha \int_0^1 [G^{-1}(u)]^k \, u^{\alpha - 1} \, du,$$

where $G^{-1}(u)$ is the baseline qf.

2.3.1.   The EE Distribution

According to Gupta and Kundu [38], if $X \sim \text{EE}(\alpha, \lambda)$,

$$\text{E}(X^k) = \frac{\alpha \, \Gamma(k+1)}{\lambda^k} \sum_{i=0}^{\infty} \frac{(-1)^i}{(i+1)^{k+1}} c(\alpha - 1, i),$$

where $c(\alpha - 1, k) = \prod_{i=1}^{k} (\alpha - i)$.

2.3.2.   The EW Distribution

According to Nadarajah *et al.* [26],

$$\text{E}(X^n) = \alpha \, c \, \lambda^c \int_0^{\infty} x^{n+\alpha-1} \, e^{-(\lambda c)^c} \left[1 - e^{-(\lambda c)^c}\right]^{\alpha-1} dx$$

and

$$\text{E}(X^n) = \alpha \, \lambda^{-n} \, \Gamma\left(\frac{n}{c} + 1\right) \sum_{i=0}^{\infty} \frac{(1-\alpha)_i}{i! \, (i+1)^{(n+c)/c}}.$$

2.4.   OTHER MOMENTS

The intensity of the populational dispersion may be measured by the deviations from the mean and median. These well-known measures are defined as the mean deviations about the mean and the median and expressed as

$$\delta_1 = \int_0^{\infty} |x - \mu_1'| \, f(x) \, dx \quad \text{and} \quad \delta_2 = \int_0^{\infty} |x - M| \, f(x) \, dx,$$

where $\mu_1' = \text{E}(X)$ and $M = \text{Median}(X)$. According to [25], these quantities can be rewritten as

$$\delta_1 = 4 \, \mu_1' \, F(\mu_1') - 2 \, m_1(\mu_1') - 2 \, \mu_1'$$

and

$$\delta_2 = \mu_1' - 2 \, m_1(\mu_1').$$

Additionally, we also present the derivation of the type-moment quantity $\text{E}(X^n | X > x)$, which is important in lifetime data contexts. In particular, the mean residual lifetime is defined by $\text{E}(X | X > x) - x$.

2.4.1.   The EE Distribution

Let $m \in \mathbb{R}^+$. Using the binomial expansion for $(1 - e^{-\lambda x})^{\alpha-1}$, Nadarajah (2011) [25] proved that

$$\int_m^\infty x\, f(x)\, dx = \frac{\alpha}{\lambda} \sum_{i=0}^\infty \frac{(-1)^i}{(i+1)^2} \binom{\alpha-1}{i} \{1 + m\lambda(i+1)\}\, e^{-m\lambda(i+1)}$$

and, from this expansion,

$$\delta_1 = 2\mu_1' F(\mu_1') - 2\mu_1'$$
$$+ \frac{2\alpha}{\lambda} \sum_{i=0}^\infty \binom{\alpha-1}{i} \frac{(-1)^i}{(i+1)^2} \{1 + \mu_1'\lambda(i+1)\}\, e^{-\mu_1'\lambda(i+1)}$$

and

$$\delta_2 = \frac{2\alpha}{\lambda} \sum_{i=0}^\infty \binom{\alpha-1}{i} \frac{(-1)^i}{(i+1)^2} \{1 + M\lambda(i+1)\}\, e^{-M\lambda(i+1)} - \mu,$$

where $\mu = \{\psi(\alpha+1) + C\}/\lambda$ and $M = (1/\lambda)\log(1 - 2^{-1/\alpha})$. For this case, Nadarajah (2011) [25] derived $E(X^n|X > x)$ in closed-form:

$$E(X^n \mid X > x) = \frac{\alpha\lambda}{\lambda^n\left[1 - (1 - e^{-\lambda x})^\alpha\right]} \frac{\partial^n}{\partial\alpha^n} B_{e^{\lambda x}}(\alpha+1, \alpha).$$

2.4.2.   The EW Distribution

Let X be a random variable having the EW distribution. By using similar arguments to those in Nadarajah (2011) [25], one can express

$$\delta_1 = 2\mu_1' F(\mu_1') - 2\mu_1' + \frac{2\alpha}{\lambda} \sum_{j=0}^\infty \frac{(-1)^j \binom{\alpha-1}{j}}{(j+1)^{1+1/c}} \Gamma\left(\frac{1}{c}+1, (j+1)\lambda^c\mu_1'^c\right)$$

and

$$\delta_2 = \frac{2\alpha}{\lambda} \sum_{j=0}^\infty \frac{(-1)^j}{(j+1)^{1+1/c}} \binom{\alpha-1}{j} \Gamma\left(\frac{1}{c}+1, (j+1)\lambda^c M^c\right) - \mu_1'.$$

For this case, Nadarajah *et al.*(2013) [26] derived $E(X^n|X > x)$ as:

$$E(X^n \mid X > x) = \frac{\alpha}{\lambda^n\left[1 - \{1 - \exp[-(\lambda x)^c]\}^\alpha\right]}$$
$$\times \sum_{j=0}^\infty \frac{(-1)^j}{(j+1)^{n/c+1}} \binom{\alpha-1}{j} \Gamma\left(\frac{n}{c}+1, (j+1)\lambda^c x^c\right).$$

2.5. INCOME MEASURES

In this section, we present four basic measures of income (Giorgi and Nadarajah, 2010 [92]): the Bonferroni curve $B(p)$, the Bonferroni index B, the Lorenz curve $L(p)$ and the Gini index G. The definitions of these measures for a random variable X with pdf $f(\cdot)$, cdf $F(\cdot)$, and qf $F^{-1}(\cdot)$, are:

$$B(p) = \frac{1}{p\mu'_1} \int_0^p F^{-1}(t)\mathrm{d}t, \quad B = 1 - \int_0^1 B(p)\mathrm{d}p,$$

$$L(p) = \frac{1}{\mu'_1} \int_0^p F^{-1}(t)\mathrm{d}t \quad \text{and} \quad G = 1 - 2\int_0^1 L(p)\mathrm{d}p,$$

respectively, where $\mu'_1 = \mathrm{E}(X)$.

2.5.1. The EE Distribution

Let $X \sim \mathrm{EE}(\alpha, \lambda)$. According to [25],

$$B(p) = \frac{\alpha/p}{\psi(\alpha+1)+C} \sum_{i=0}^{\infty} \frac{(-1)^i}{(i+1)^2} \binom{\alpha-1}{i}$$
$$\times \left\{ 1 - \left[1 - (i+1)\log(1-p^{1/\alpha}) \right] (1-p^{1/\alpha})^{i+1} \right\},$$

$$B = 1 - \frac{\alpha^2}{\psi(\alpha+1)+C} \sum_{i=0}^{\infty} \frac{(-1)^i}{(i+1)^2} \binom{\alpha-1}{i}$$
$$\times \left[\sum_{j=1}^{\infty} \binom{i+1}{j} \frac{(-1)^{j+1}}{j} - (i+1)\,\psi'(i+3) \right],$$

$$L(p) = \frac{\alpha}{\psi(\alpha+1)+C} \sum_{i=0}^{\infty} \frac{(-1)^i}{(i+1)^2} \binom{\alpha-1}{i}$$
$$\times \left\{ 1 - \left[1 - (i+1)\log(1-p^{1/\alpha}) \right] (1-p^{1/\alpha})^{i+1} \right\}$$

and

$$G = 1 - \frac{2\,\alpha}{\psi(\alpha+1)+C} \sum_{i=0}^{\infty} \frac{(-1)^i}{(i+1)^2} \binom{\alpha-1}{i} \left\{ 1 - \alpha\,B(\alpha, i+2) \right.$$
$$\left. + (i+1)\,B(\alpha, i+2)\left[\psi(i+2) - \psi(\alpha+i+2) \right] \right\},$$

where C is Euler's constant.

2.5.2. The EW Distribution

Letting $X \sim \text{EW}(c, \alpha, \lambda)$, Sarabia and Castillo (2005) [93] proved that

$$L(u) = \frac{\int_0^{u^{1/\alpha}} y^{\alpha-1} F_0^{-1}(y)\, dy}{\int_0^1 y^{\alpha-1} F_0^{-1}(y)}\, dy$$

and

$$G = \frac{1}{E(X)} \int_0^1 \frac{u^\alpha - u^{2\alpha}}{f_0\left(F_0^{-1}(u)\right)}\, du,$$

where f_0 and F_0 are the pdf and cdf of a two-parameter Weibull distribution with shape parameter c and scale parameter λ.

Here, we provide explicit expressions for four income measures for the EW distribution. By using the power series for $\log(1-z)$, one can rewrite (2.9)

$$\begin{aligned}
F^{-1}(p) &= \frac{1}{\lambda}\left[\sum_{k=1}^\infty \frac{p^{k/\alpha}}{k}\right]^{1/c} \\
&= \frac{p^{1/(\alpha c)}}{\lambda}\left[1 + \sum_{k=2}^\infty \frac{p^{k/\alpha}}{k}\right]^{1/c} \\
&= \frac{p^{1/(\alpha c)}}{\lambda}\sum_{i=0}^\infty \binom{1/c}{i}\left[\sum_{k=2}^\infty \frac{p^{k/\alpha}}{k}\right]^i \\
&= \frac{1}{\lambda}\sum_{i=0}^\infty \binom{1/c}{i}\sum_i \frac{p^{(k_1+\cdots+k_i-i-1/c)/\alpha}}{k_1\cdots k_i},
\end{aligned}$$

where $\sum_i$ denotes summation over $k_1 \geq 2, \ldots, k_i \geq 2$. Using this representation, we can easily calculate the four measures:

$$B(p) = \frac{1}{\lambda\mu_1'}\sum_{i=0}^\infty \binom{1/c}{i}\sum_i \frac{p^{(k_1+\cdots+k_i-i-1/c)/\alpha}}{k_1\cdots k_i\left[(k_1+\cdots+k_i-i-1/c)/\alpha+1\right]},$$

$$B = 1 - \frac{1}{\lambda\mu_1'}\sum_{i=0}^\infty \binom{1/c}{i}\sum_i \frac{1}{k_1\cdots k_i\left[(k_1+\cdots+k_i-i-1/c)/\alpha+1\right]^2},$$

$$L(p) = \frac{1}{\lambda\mu_1'}\sum_{i=0}^\infty \binom{1/c}{i}\sum_i \frac{p^{(k_1+\cdots+k_i-i-1/c)/\alpha}+1}{k_1\cdots k_i\left[(k_1+\cdots+k_i-i-1/c)/\alpha+1\right]}$$

and

$$G = 1 - \frac{2}{\lambda \mu_1'} \sum_{i=0}^{\infty} \binom{1/c}{i} \sum_i \frac{\left[(k_1 + \cdots + k_i - i - 1/c) / \alpha + 2 \right]^{-1}}{k_1 \cdots k_i \left[(k_1 + \cdots + k_i - i - 1/c) / \alpha + 1 \right]}.$$

2.6.   ORDER STATISTICS

Let $X_1, \ldots, X_n$ be a random sample from X having cdf and pdf given by $F(x)$ and $f(x)$, respectively. Consider $X_{1:n} < X_{2:n} < \cdots < X_{n:n}$ the ordered observations from a random sample of size n such that $X_{r:n}$ is the rth order statistic from the n-dimensional sample. Thus, the cdf and pdf of $Y = X_{r:n}$ are given by

$$F_Y(y) = \sum_{j=r}^{n} \binom{n}{j} F(x)^j \left[1 - F(x) \right]^{n-j}$$

$$= \sum_{j=r}^{n} \binom{n}{j} G(x)^{\alpha j} \left[1 - G(x)^\alpha \right]^{n-j}$$

and

$$f_Y(x) = B(r, n-r+1)^{-1} f(x) F(x)^{r-1} [1 - F(x)]^{n-r}$$
$$= \alpha B(r, n-r+1)^{-1} g(x) G(x)^{\alpha r - 1} [1 - G(x)^\alpha]^{n-r},$$

respectively. Further, $X_{1:n}$ and $X_{n:n}$ are called the *extreme order statistics*, and $R = X_{n:n} - X_{1:n}$ is called the *extreme spacing*.
Thus, setting $F^\alpha(x) = u$, one has

$$\mathrm{E}(Y^k) = \mathrm{E}(X_{r:n}^k) = B(r, n-r+1)^{-1} \int_0^1 u^{r-1} (1-u)^{n-r} \left[F^{-1}(u^{1/\alpha}) \right]^k \mathrm{d}u.$$

Subsequently, we discuss two cases:

2.6.1.   For the EE Distribution

According to Nadarajah (2011) [25], assuming $X \sim \mathrm{EE}(\alpha, \lambda)$ with $\lambda = 1$,

$$f_{r:n}(x) = \sum_{j=0}^{n-r} d_j(n, r) f(x; \alpha_{r+j}),$$

where $\alpha_j = j \alpha$, $d_j(n, r) = (-1)^j n \binom{n-1}{r-1} \binom{n-r}{j} \big/ (r+j)$ and $f(x; \alpha_{r+j})$ denotes the EE model with $(r+j)\alpha$ and 1 as shape and scale parameters, respectively. In particular,

$$f_{n:n}(x; \alpha) = n \alpha \, \mathrm{e}^{-x} (1 - \mathrm{e}^{-x})^{n\alpha - 1}.$$

Using binomial and power series expansions and, after some algebra, we obtain (for $|t| < 1$)

$$M_{X_{r:n}}(t) = \mathrm{E}(e^{t\,X_{r:n}}) = \alpha\,B(r, n - r + 1)^{-1}$$
$$\sum_{j=0}^{n-r}(-1)^j \binom{n-r}{j} \frac{\Gamma(\alpha(r+j))\Gamma(1-t)}{\Gamma(\alpha(r+j)-t+1)}.$$

By differentiating $M_{X_{r:n}}(t)$ twice in t and evaluating at $t = 0$, we obtain

$$\mathrm{E}(X_{r:n}) = B(r, n - r + 1)^{-1} \sum_{j=0}^{n-r}(-1)^j \frac{\binom{n-r}{j}}{(r+j)} \left[\psi(\alpha(r+j)+1) - \psi(1)\right]$$

and

$$\mathrm{E}(X_{r:n}^2) = B(r, n - r + 1)^{-1} \sum_{j=0}^{n-r}(-1)^j \frac{\binom{n-r}{j}}{(r+j)}$$
$$\times \left\{\left[\psi(\alpha(r+j)+1) - \psi(1)\right]^2 + \psi'(1) - \psi'(\alpha(r+j)+1)\right\}.$$

2.6.2.   For the EW Distribution

Mudholkar and Hutson (1996) [59] derived asymptotic distributions of the extreme order statistics, $X_{1:n}$ and $X_{n:n}$, and the extreme spacings, $X_{2:n} - X_{1:n}$ and $X_{n:n} - X_{n-1:n}$. They proved that $n^{1/(\alpha c)} X_{1:n}$ approaches in distribution $Z^{1/(\alpha c)}$ such that Z is a standard exponential random variable. Further,

$$n^{1/(\alpha c)}\left(X_{2:n} - X_{1:n}\right) \to Z^{1/(\alpha c)} - X^{1/(\alpha c)}$$

and

$$(\log n)^{1-1/c}\left[X_{n:n} - X_{n-1:n}\right] \to (\log Z - \log X)/c,$$

where (Z, X) has the joint pdf

$$f(z, x) = \begin{cases} e^{-z}, & \text{if } 0 \le x \le z, \\ 0 & \text{otherwise.} \end{cases}$$

Further, $X_{2:n} - X_{1:n} = O_p(n^{-1/(\alpha c)})$ and $X_{n:n} - X_{n-1:n} = O_p((\log n)^{1/c-1})$. Sarabia and Castillo (2005) [93] demonstrated that the pdf of the rth order statistic, say $Y = X_{r:n}$, can be expressed as

$$f_Y(y) = \sum_{j=0}^{n-r} \frac{(-1)^j\, n!}{(r-1)!\,(n-r-j)!\,(r+j)!}\, f_{(r+j)\,\alpha}(y),$$

where $f_{(r+j)\alpha}(y)$ is given by (2.8) with α replaced by $(r+j)\alpha$. Thus, the pdf of the rth order statistic is a linear combination of pdfs of the form (2.8). Hence, the corresponding cdf, survival function, moments and product moments can be easily determined. Raqab and Kundu (2006) [94] derived an expression for a product moment of order statistics.

Raqab (1998) [95] determined explicit expressions for single and product moments of order statistics from the EW distribution when $c = 2$ and $\lambda = 1$. He also obtained expressions for percentiles and proposed an estimate for α based on order statistics. Khan *et al.* (2008) [96] established recurrence expressions based on the EW law for generalized order statistics.

2.7.   ENTROPY

An entropy of a random variable X is a measure of variation of the uncertainty. Three popular measures of entropy are:

- the Rényi entropy (Rényi, 1961 [97]) defined by

$$\mathcal{J}_R(\gamma) = \frac{1}{(1-\gamma)} \log \left(\int f^\gamma(x)\,\mathrm{d}x \right),\qquad (2.13)$$

  where $\gamma > 0$ and $\gamma \neq 1$;

- the Shannon entropy (Shannon, 1948 [19]) defined by $\mathrm{E}[-\log f(X)]$, the particular case of (2.13) when $\gamma \uparrow 1$; and,

- the cumulative residual entropy (Rao *et al.*, 2004 [98]) defined by

$$\mathcal{J}_C = -\int \Pr(X > x)\, \log[\Pr(X > x)]dx.\qquad (2.14)$$

In this section, we present expressions for these three entropies when X is a random variable having the pdfs (2.4) and (2.8).

2.7.1.   The EE Distribution

Let $X \sim \mathrm{EE}(\alpha, \lambda)$. Then, the following results hold:

$$\mathcal{J}_R(\gamma) = -\log(\alpha\,\lambda) + \frac{\log(\alpha) + \log(B(\gamma, \gamma\alpha - \gamma + 1))}{1-\gamma},$$

$$\mathcal{J}_R(1) = -\log(\alpha\,\lambda) - \psi(1) + (1-\alpha)\,\psi(\alpha) + \alpha\,\psi(\alpha+1),$$

and

$$\mathcal{J}_C = \frac{1}{\lambda} \sum_{j,k=1}^{\infty} \binom{\alpha}{j} B(j, \alpha\,k + 1).$$

2.7.2.   The EW Distribution

First, consider the Rényi entropy. Using arguments similar to those in Nadarajah (2011, Section 5) [25], one can express (2.13) as

$$\mathcal{J}_R(\gamma) = \frac{\gamma}{(1-\gamma)} \log\left(\frac{\alpha}{\gamma}\right) - \log\left(c\,\lambda\,\gamma^{1/c}\right)$$

$$+ \frac{1}{(1-\gamma)} \log\left\{ \sum_{j=0}^{\infty} \frac{(-1)^j \binom{\alpha\gamma-\gamma}{j}}{(j+1)^{(1/c-1)(1-\gamma)}} \Gamma\left((1/c-1)(1-\gamma)+1\right) \right\}.$$

The Shannon entropy can be obtained by limiting $\gamma \uparrow 1$ in the last equation. However, it is easier to derive an expression for it from first principles. Using the power series for $\log(1-z)$, we obtain

$$
\begin{aligned}
E\left[-\log f(X)\right] &= -\log\left(c\alpha\lambda^c\right) + (1-c)E\left(\log X\right) + \lambda^c E\left(X^c\right) \\
&\quad + (\alpha-1)\sum_{k=1}^{\infty}\frac{1}{k}E\left[\exp\left\{-k(\lambda X)^c\right\}\right] \\
&= -\log\left(c\alpha\lambda^c\right) + (1-c)E\left(\log X\right) - \psi(1) + \psi(\alpha+1) \\
&\quad + \alpha(\alpha-1)\sum_{k=1}^{\infty}\frac{1}{k}B\left(k+1,\alpha\right).
\end{aligned}
$$

The computation of $E[\log(X)]$ above requires numerical software. Finally, consider the cumulative residual entropy. Using the power series for $\log(1-z)$, and arguments similar to those in Nadarajah (2011, Section 5) [25], one can express (2.14) as

$$\mathcal{J}_C = \frac{\Gamma(1/c)}{\lambda c} \sum_{i,j=1}^{\infty} \sum_{k=0}^{\infty} \binom{\alpha}{i}\binom{j\alpha}{k} \frac{(-1)^{i+k}}{j\,(i+k)^{1/c}}.$$

2.8.   ESTIMATION

Consider a random variable having the exp-G distribution and let $\boldsymbol{\theta} = (\alpha, \boldsymbol{\delta}^\top)^\top$ be its parameter vector. The log-likelihood for $\boldsymbol{\theta}$ based on any observation x is

$$\ell(\boldsymbol{\theta};x) = \log(\alpha) + \log[g(x;\boldsymbol{\delta})] + (\alpha-1)\log[G(x;\boldsymbol{\delta})]. \qquad (2.15)$$

The MLE of $\boldsymbol{\theta}$ is determined by maximizing $\ell_n(\boldsymbol{\theta}) = \sum_{i=1}^{n} \ell(\boldsymbol{\theta};x_i)$ from a n-dimensional data set $x_1,\ldots,x_n$.

Based on (2.15), the score function

$$\boldsymbol{U}_\theta = (U_\alpha, U_\delta^\top)^\top = \left(\frac{\mathrm{d}\ell_n(\boldsymbol{\theta})}{\mathrm{d}\alpha}, \left[\frac{\mathrm{d}\ell_n(\boldsymbol{\theta})}{\mathrm{d}\boldsymbol{\delta}} \right]^\top \right)^\top,$$

can be derived as follows.
After some algebra, two identities hold:

$$U_\alpha = \frac{n}{\alpha} + \sum_{i=1}^{n} \log[G(x_i; \boldsymbol{\delta})] \tag{2.16}$$

and

$$U_\delta = \sum_{i=1}^{n} g(x_i; \boldsymbol{\delta})^{-1} \left[\frac{\partial g(x_i; \boldsymbol{\delta})}{\partial \boldsymbol{\delta}} \right] + (\alpha - 1) \sum_{i=1}^{n} G(x_i; \boldsymbol{\delta})^{-1} \left[\frac{\partial G(x_i; \boldsymbol{\delta})}{\partial \boldsymbol{\delta}} \right]. \tag{2.17}$$

From the first identity and given the MLE, $\widehat{\boldsymbol{\delta}}$, the MLE $\widehat{\alpha}$ has partially the closed-form given by

$$\widehat{\alpha} = \frac{-n}{\sum_{i=1}^{n} \log[G(x_i; \widehat{\boldsymbol{\delta}})]}.$$

Finally, the partitioned observed information matrix for the exp-G class is

$$J(\boldsymbol{\theta}) = \begin{pmatrix} U_{\alpha\alpha} & | & U_{\alpha\delta^\top} \\ -- & & -- \\ U_{\delta\alpha} & | & U_{\delta^\top\delta} \end{pmatrix} = \begin{pmatrix} \frac{\mathrm{d}^2 \ell_n(\boldsymbol{\theta})}{\mathrm{d}\alpha^2} & | & \frac{\mathrm{d}^2 \ell_n(\boldsymbol{\theta})}{\mathrm{d}\alpha \, \mathrm{d}\boldsymbol{\delta}^\top} \\ -- & & -- \\ \frac{\mathrm{d}^2 \ell_n(\boldsymbol{\theta})}{\mathrm{d}\boldsymbol{\delta}\mathrm{d}\alpha} & | & \frac{\mathrm{d}^2 \ell_n(\boldsymbol{\theta})}{\mathrm{d}\boldsymbol{\delta}^\top \mathrm{d}\boldsymbol{\delta}} \end{pmatrix},$$

whose elements are

$$U_{\alpha\alpha} = -\frac{n}{\alpha^2}, \quad U_{\delta\alpha} = \sum_{i=1}^{n} G(x_i)^{-1} \left[\frac{\partial G(x_i)}{\partial \boldsymbol{\delta}} \right]$$

and

$$\begin{aligned} U_{\delta^\top\delta} &= -\sum_{i=1}^{n} g(x_i)^{-2} \left[\frac{\partial g(x_i)}{\partial \boldsymbol{\delta}} \right] \left[\frac{\partial g(x_i)}{\partial \boldsymbol{\delta}^\top} \right] + \sum_{i=1}^{n} g(x_i)^{-1} \left[\frac{\partial^2 g(x_i)}{\partial \boldsymbol{\delta}^\top \partial \boldsymbol{\delta}} \right] \\ &= -(\alpha - 1) \sum_{i=1}^{n} G(x_i)^{-2} \left[\frac{\partial G(x_i)}{\partial \boldsymbol{\delta}} \right] \left[\frac{\partial G(x_i)}{\partial \boldsymbol{\delta}^\top} \right] \\ &\quad + \sum_{i=1}^{n} G(x_i)^{-1} \left[\frac{\partial^2 G(x_i)}{\partial \boldsymbol{\delta}^\top \partial \boldsymbol{\delta}} \right]. \end{aligned}$$

2.8.1. For the EE Distribution

Let $x_1, \ldots, x_n$ be a sample drawn from $X \sim \text{EE}(\alpha, \lambda)$. From (2.16) and (2.17), the MLE of λ is given by a solution of

$$\widehat{\lambda} = \left[\frac{\sum_{i=1}^{n} x_i e^{-\widehat{\lambda} x_i} \big/ (1 - e^{-\widehat{\lambda} x_i})}{\sum_{i=1}^{n} \log(1 - e^{-\widehat{\lambda} x_i})} - \frac{1}{n} \sum_{i=1}^{n} \frac{x_i}{e^{-\widehat{\lambda} x_i}} \right]^{-1}.$$

Given $\widehat{\lambda}$, the MLE $\widehat{\alpha}$ has the form

$$\widehat{\alpha}(\widehat{\lambda}) = -\frac{n}{\sum_{i=1}^{n} \log(1 - e^{-\widehat{\lambda}})}.$$

Finally, as the EE family satisfies general regularity conditions [44], the following result holds:

$$\sqrt{n} \begin{bmatrix} \widehat{\alpha} - \alpha \\ \widehat{\lambda} - \lambda \end{bmatrix} \xrightarrow[n \to \infty]{\mathcal{D}} N_2(\mathbf{0}, \mathbf{\Sigma}),$$

where $\mathbf{\Sigma}$ is given in [43].

2.8.2. For the EW Distribution

Mudholkar and Srivastava (1993) [7] proved that the MLEs of (c, λ, α) are the simultaneous solutions of

$$\frac{n}{c} + (\alpha - 1) \sum_{i=1}^{n} \frac{g_1(x_i)}{g(x_i)} - \sum_{i=1}^{n} (\lambda x_i)^c \log(\lambda x_i) + \sum_{i=1}^{n} \log(\lambda x_i) = 0,$$

$$\frac{n}{\alpha} + \sum_{i=1}^{n} \log g(x_i) = 0,$$

$$-nc\lambda + (\alpha - 1) \sum_{i=1}^{n} \frac{g_2(x_i)}{g(x_i)} + \lambda c \sum_{i=1}^{n} (\lambda x_i)^c = 0,$$

where $g(x) = 1 - \exp\left\{-(\lambda x)^c\right\}$, $g_1(x) = (\lambda x)^c \log(\lambda x) \exp\left\{-(\lambda x)^c\right\}$, and $g_2(x) = -\lambda c (\lambda x)^c \exp\left\{-(\lambda x)^c\right\}$. In practice, the solution for the above nonlinear equations system can be found by a numerical optimization method such as the Broyden–Fletcher–Goldfarb–Shanno (BFGS) algorithm.

Cheng *et al.* (2010) [99] proposed an EM algorithm for computing the MLEs. But they found the EM algorithm ineffective, so they suggested a new algorithm to improve effectiveness. Pal *et al.* (2006) [61] investigated some analytical properties of the surface for the EW log-likelihood. They proved, for given (λ, c), that the log-likelihood is a strictly concave function of α. Also, the

MLE of α is a concave increasing function of λ for given c. Further, for given (α, c), and $\alpha \geq 1$, the log-likelihood is a strictly concave function of λ.

By asymptotic normality, $(\widehat{\alpha} - \alpha, \widehat{c} - c, \widehat{\lambda} - \lambda)$ approaches a trivariate normal random vector with zero means and covariance matrix $\mathbf{K}(\boldsymbol{\theta})^{-1}$, where

$$
\frac{1}{n}K(\boldsymbol{\theta}) = \left[
\begin{array}{c:cc}
\mathrm{E}\left(-\dfrac{\partial^2 \log L}{\partial \alpha^2}\right) & \mathrm{E}\left(-\dfrac{\partial^2 \log L}{\partial \alpha \partial c}\right) & \mathrm{E}\left(-\dfrac{\partial^2 \log L}{\partial \alpha \partial \lambda}\right) \\
\hdashline
\mathrm{E}\left(-\dfrac{\partial^2 \log L}{\partial c \partial \alpha}\right) & \mathrm{E}\left(-\dfrac{\partial^2 \log L}{\partial c^2}\right) & \mathrm{E}\left(-\dfrac{\partial^2 \log L}{\partial c \partial \lambda}\right) \\
\mathrm{E}\left(-\dfrac{\partial^2 \log L}{\partial \lambda \partial \alpha}\right) & \mathrm{E}\left(-\dfrac{\partial^2 \log L}{\partial \lambda \partial c}\right) & \mathrm{E}\left(-\dfrac{\partial^2 \log L}{\partial \lambda^2}\right)
\end{array}
\right].
$$

For interval estimation and hypothesis tests, it is useful to have explicit expressions for the elements of $K(\boldsymbol{\theta})$, the expected unit information matrix. From the result of Cordeiro *et al.* (2011c) [90], we can write

$$
\kappa_{11} = \psi'(\alpha) - \psi'(\alpha + 1), \ \kappa_{12} = -\frac{\alpha}{c}T(2, 2, \alpha - 1, 1),
$$

$$
\kappa_{13} = -\frac{c\alpha}{\lambda}T(2, 2, \alpha - 1, 0),
$$

$$
\kappa_{22} = \frac{1}{c^2} + \frac{\alpha}{c^2}\Big[(\alpha - 1)T(3, 2, \alpha - 2, 2) + T(2, 1, \alpha - 2, 2)
$$
$$
- (\alpha + 1)T(2, 2, \alpha - 2, 2) + \alpha T(2, 3, \alpha - 2, 2)\Big],
$$

$$
\kappa_{23} = \frac{\alpha}{\lambda}\Big[(\alpha - 1)T(3, 2, \alpha - 2, 1) + T(2, 1, \alpha - 2, 1)
$$
$$
- (\alpha + 2)T(2, 2, \alpha - 2, 1) + \alpha T(2, 3, \alpha - 2, 1)\Big],
$$

$$
\kappa_{33} = \frac{c^2}{\lambda^2} + \frac{c^2 \alpha(\alpha - 1)}{\lambda^2}T(3, 2, \alpha - 2, 0),
$$

where $T(d, b, a, c) = B(a, b)\,\mathrm{E}[W^{d-1}\log^c W]$, $W = -\log(U)$ and U is a beta random variable with shape parameters (b, a).

2.9. APPLICATION

As an example of application for the exponentiated models, we use a real data set to compare the fits of the EW distribution with two sub-models, such as the EE and Weibull distributions. For each model, the parameters are estimated by maximum likelihood using the `NLMixed` subroutine in `SAS`.

A database on fatigue life of 6061-T6 aluminum coupons has been discussed by Birnbaum and Saunders (1969) [100]. It was composed by 101 measurements with at most 31,000 psi stress per cycle. In Table 2.1, we list the MLEs of the parameters (with corresponding standard errors in parentheses) and the AIC and BIC statistics for the EW, EE and Weibull models.

Table 2.1: Parameter estimates, AIC and BIC statistics for the Birnbaum and Saunders' data.

Model	Estimates			AIC	BIC
$EW(c, \alpha, \lambda)$	5.7105	0.0098	2.9712	911.0626	918.8781
	(2.2489)	(0.0027)	(0.1292)		
$EE(\alpha, \lambda)$	119.3648	0.0391		926.1762	931.3866
	(29.5198)	(0.0380)			
$Weibull(\alpha, \beta)$	0.0070	5.9790		922.1997	927.4100
	(0.0013)	(2.2381)			

Roughly, we conclude that all competing distributions can be used to model these data. However, the fitted EW, EE and Weibull densities are displayed in Figure 2.1. They indicate that the EW distribution yields a better fit to these data than the EE and Weibull distributions. These conclusions are confirmed from the values of the AIC and BIC statistics for the fitted models given in Table 2.1.

2.10. CONCLUSIONS

The statistics literature has numerous distributions for modeling lifetime data. But most of them lack motivation from a lifetime context. For example, there is no apparent physical motivation for the gamma and Weibull models and many other distributions. Many if not most lifetime distributions are motivated only by mathematical interest. The exponentiated G family is an extension of a baseline G distribution obtained by adding an extra power shape parameter, raising the parent cdf to a positive power. Various special exponentiated models are presented. We derive some mathematical properties of the exponentiated exponential and exponentiated Weibull models. In particular, we obtain explicit expressions for the ordinary and incomplete moments, quantile and generating functions, income measures, entropies and order statistics. The parameters of the exponentiated family are estimated by maximum likelihood. We illustrate its flexibility by means of a real data set.

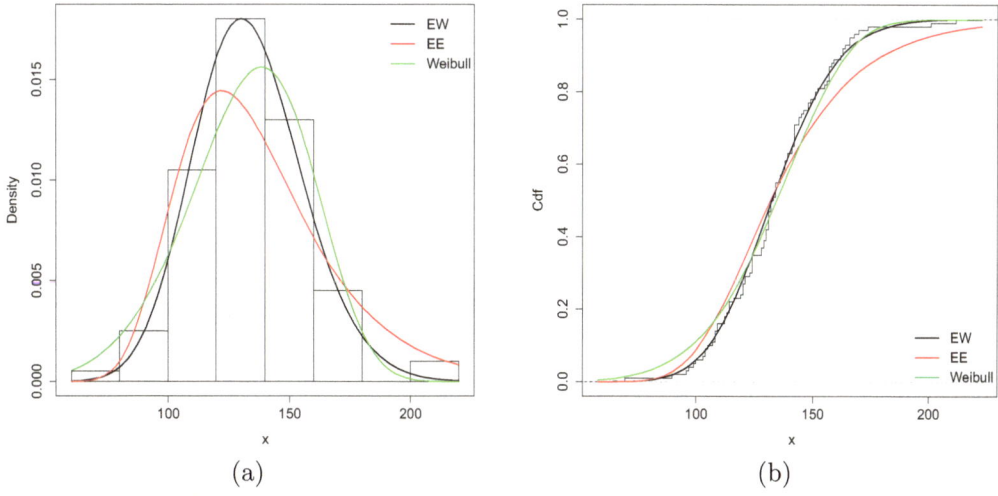

Figure 2.1: Estimated (a) pdf and (b) cdf for the EW, EE and Weibull models for the current data.

Chapter 3

Beta Generalized Models

Abstract: The beta transformation gives a great variety of shapes which allow to model symmetric, skewed and bimodal shaped densities. Such procedure generates what we call the beta generalized (beta-G) family of distributions. This chapter presents a survey on the beta-G models and their mathematical properties. We present some explicit expressions for the ordinary and incomplete moments, probability weighted moments, cumulants and generating function, mean deviations, entropies and order statistics.

Keywords:: Beta-G model; Entropy; Mean deviation; Moment; Probability weighted moment; Regression model.

3.1. INTRODUCTION

Several distributions have been employed in order to perform inference on populational properties from one or more observed samples. Choosing the model which should be adopted to test hypothesis about the data is a crucial step in statistical data analysis. In this chapter, the *beta-G* ("BG") family of distributions proposed by Eugene *et al.* (2002) [101] is studied in details. This family includes nearly all of well-known models as special or limiting cases such as those exponentiated distributions. Further, it can give lighter tails and heavier tails and be applied in some areas such as engineering and biological research, among others. Explicit expressions are reported, which facilitate to obtain several mathematical properties of this family.

In the last years, several BG models have been proposed in this family, mostly by statisticians in Brazil. This family has the major benefit for fitting skewed data that can not be fitted by most well-known continuous distributions. Starting from a baseline cdf $G(x; \boldsymbol{\tau})$, where $\boldsymbol{\tau}$ indicates the parameters of

Gauss M. Cordeiro, Rodrigo B. Silva & Abraão D. C. Nascimento

the cdf $G(\cdot)$, Eugene *et al.* (2002) [100] defined the BG family by the cdf (for $x \in \mathbb{R}$)

$$F(x; a, b, \boldsymbol{\tau}) = I_{G(x;\boldsymbol{\tau})}(a, b) = \frac{1}{B(a,b)} \int_0^{G(x;\boldsymbol{\tau})} \omega^{a-1}(1-\omega)^{b-1} d\omega, \qquad (3.1)$$

where $a > 0$ and $b > 0$ are shape parameters, $I_y(a, b) = B_y(a, b)/B(a, b)$ is the incomplete beta function ratio, $B_y(a, b) = \int_0^y \omega^{a-1}(1-\omega)^{b-1} d\omega$ is the incomplete beta function, $B(a, b) = \Gamma(a)\Gamma(b)/\Gamma(a+b)$ is the beta function and $\Gamma(a) = \int_0^\infty \omega^{a-1}e^{-\omega} d\omega$ is the gamma function. If $G(x; \boldsymbol{\tau}) = x$ for $x \in (0, 1)$, we obtain the beta distribution. The pdf associated with (3.1) can be written as (for $x \in \mathbb{R}$)

$$f(x; a, b, \boldsymbol{\tau}) = \frac{1}{B(a,b)} g(x; \boldsymbol{\tau}) G(x; \boldsymbol{\tau})^{a-1} [1 - G(x; \boldsymbol{\tau})]^{b-1}, \qquad (3.2)$$

where $g(x; \boldsymbol{\tau}) = dG(x; \boldsymbol{\tau})/dx$ is the baseline pdf. The manageability of $f(x; a, b, \boldsymbol{\tau})$ is linked with the forms of $G(x; \boldsymbol{\tau})$ and $g(x; \boldsymbol{\tau})$. In fact, depending on the complexity of these functions, we can take considerable time and effort to work with the density (3.2) in generality.

If G has the support in $\mathbb{R}^+$, the BG hrf has the form

$$h(x; a, b, \boldsymbol{\tau}) = \frac{g(x; \boldsymbol{\tau}) \, G(x; \boldsymbol{\tau})^{a-1} \, [1 - G(x; \boldsymbol{\tau})]^{b-1}}{B(a, b) \, [1 - I_{G(x;\boldsymbol{\tau})}(a, b)]}.$$

We use $G(x)$ instead of $G(x; \boldsymbol{\tau})$, $g(x)$ instead of $g(x; \boldsymbol{\tau})$, $F(x)$ instead of $F(x; a, b, \boldsymbol{\tau})$, *etc*, to simplify the notation.

Throughout this chapter, the random variable X having density function (3.2) is denoted by $X \sim \text{BG}(a, b, \boldsymbol{\tau})$. It can be expressed by the stochastic representation $X = Q(U) = F^{-1}(U)$, where $U \sim \text{Beta}(a, b)$ and $Q(\cdot)$ is the inverse function of (3.1). Further, we can write (3.1) in terms of the Gaussian hypergeometric function (Gradshteyn and Ryzhik, 2000; Section 9.1 [11]). The properties of this function are well-known. We have

$$F(x) = \frac{G(x)}{a \, B(a, b)} \, {}_2F_1(a, 1 - b; a + 1; G(x)), \quad x \in \mathbb{R},$$

where

$$ {}_2F_1(a, b; c; z) = \frac{\Gamma(c)}{\Gamma(a)\,\Gamma(b)} \sum_{j=0}^{\infty} \frac{\Gamma(a+j)\,\Gamma(b+j)}{\Gamma(c+j)} \frac{z^j}{j!}.$$

One important special model of the BG family is the exp-G class, discussed in Chapter 2, which arises when $b = 1$ in (3.2). The BG family received great consideration in the last years, after the proposals of Eugene *et al.* (2002) [100] and Jones (2004) [102]. After these seminal works, many extended models were introduced and studied. Gupta and Nadarajah (2004) [103] determined a more general formula for the *n*th moment of the beta normal (BN) distribution. Razzaghi (2009) [104] adopted the BN distribution for risk assessment and to model dose-response, where the BN properties are discussed and the risk estimates are based upon the asymptotic properties of the MLEs. Recently, Rêgo *et al.* (2012) [105] furnished a better treatment for the BN distribution, derivingseveral of its properties and a detailed discussion on its bimodality. They derived a power series for the qf to obtain computable expressions for the moments, generating function and mean deviations. Further, the BN law has been employed successfully to synthetic aperture radar imagery processing; see Cintra *et al.* (2012) [106]. fgumbelThese authors proposed the beta generalized normal distribution by compounding the beta and generalized normal distributions.

The first distribution of the BG class was the BN model (Eugene *et al.*, 2002 [100]). Denote the standard normal cdf and pdf by $\Phi(\cdot)$ and $\phi(\cdot)$, respectively. Let $X = \Phi^{-1}(U)$, where $U \sim \text{Beta}(a, b)$. Then, X has a standard BN distribution, say $\text{BN}(a, b, 0, 1)$, if its pdf has the form

$$f(x; a, b, 0, 1) = \frac{1}{B(a,b)} \, \phi(x) \, \Phi(x)^{a-1} \, [1 - \Phi(x)]^{b-1}, \quad x \in \mathbb{R}.$$

The skewness and kurtosis of X usually depend on the extra parameters a and b, see Table 3.1. Eugene *et al.* (2002) evaluated the first four cumulants of X with $\mu = 0$ and $\sigma^2 = 1$ for some values of these parameters between 0.05 and 100. The skewness of X is in the interval $(-1, 1)$ and the largest kurtosis value is 4.1825 for $a = 100$ and $b = 0.1$. If $X \sim \text{BN}(a, b, 0, 1)$, then $Y = \sigma X + \mu \sim \text{BN}(a, b, \mu, \sigma)$ has the non-standard BN distribution with parent $\text{N}(\mu, \sigma^2)$.

New distributions in the BG family were investigated in the last ten years. Some of them are now described in the order they were published. The beta Fréchet (BFr) distribution follows from the Fréchet cdf $G(x)$. It was defined by Nadarajah and Gupta (2004) [103], who studied analytically its density and hrf as well as the limit distribution of the order statistics. Nadarajah and Kotz (2004) [41] proposed the beta Gumbel distribution from the Gumbel cdf $G(x)$ and yielded expressions for its moments and the asymptotic distribution of the order statistics. A generalization of the Weibull model called the beta Weibull (BW) distribution was presented by Famoye *et al.* (2005) [80].

Gupta and Nadarajah (2006) [106] introduced the beta Bessel distribution

Table 3.1: Skewness and kurtosis of the BN distribution for some parameters values.

Parameters values	skewness
$a = b$	symmetric
$a < b$	negative
$a > b$	positive

Parameters values	kurtosis
$a = b > 1$	positive excess
$a = b < 1$	negative excess

and studied its shapes. The moment, Bayesian and maximum likelihood methods were addressed by these authors. Nadarajah and Kotz (2006) [41] pioneered the beta exponential (BE) distribution and derived the first four moments, mgf and discussed maximum likelihood estimation. Rodríguez-Avi *et al.* (2007) [107] proposed a generalization of the beta-binomial distribution generated by the Gaussian hypergeometric function. Lee *et al.* (2007) [81] investigated the BW distribution for censored data and proved that its hrf can have unimodal, bathtub, increasing and decreasing shapes.

Akinsete *et al.* (2008) [108] introduced a four-parameter beta-Pareto (BPa) distribution. Various of its properties were derived and it was applied to two flood data sets. Kozubowski and Nadarajah (2008) [109] introduced the beta Laplace (BLa) distribution and derived structural properties, including modality and concavity of the density, moments, and stochastic representations that aid in random variate generation from this model.

Further, the beta modified Weibull (BMW) distribution was proposed by Silva *et al.* (2010b) [85] for analysis of lifetime data. Their model can be used in the survival analysis since its hrf can be monotone, unimodal and bathtub-shaped. Barreto–Souza *et al.* (2010) [34] studied the beta generalized exponential distribution. Pescim *et al.* (2010) [110] introduced the beta generalized half-normal model, which extends the generalized half-normal and half-normal distributions. Barreto–Souza *et al.* (2011) [111] determined the ordinary and L-moments of the BFr distribution. They also discussed the estimation of the model parameters and obtained the information matrix.

Cordeiro *et al.* (2011a) [90] obtained explicit expressions for the moments of the BW distribution, by extending well-known results available for some sub-models. The beta generalized Pareto distribution is an extension of the BMW model defined by Mahmoudi (2011) [112]. The author proved that it is more flexible than the BPa distribution and has some interesting properties. Nadarajah *et al.* (2011) [113] obtained explicit expressions for the BMW distribution such as the moments, mgf, mean deviations, reliability and entropies.

Cordeiro and Lemonte (2011) [114] introduced the BLa distribution, determined some of its mathematical properties and investigated maximum likelihood estimation for the model parameters. Further, Cordeiro and Lemonte (2011b) [115] presented the beta half-Cauchy (BHC) distribution and proved empirically that it can produce a better fit than the Birnbaum-Saunders, gamma and Weibull models by means of a real data set.

Castellares *et al.* (2011) [116] introduced the beta log-normal (BLN) distribution and Cordeiro and Nadarajah (2011) [82] obtained closed-form expressions for the moments of the beta generalized distributions. Paranaíba *et al.* (2011) [117] presented the five-parameter beta Burr XII (BBXII) distribution, which has as special models the logistic, Weibull and Burr XII distributions, among others. The authors also derived its mgf. Cordeiro and Lemonte (2011c) [118] pioneered the beta-Birnbaum-Saunders (BBS) distribution for fatigue life modeling and obtained some of its properties.

Recently, Alexander *et al.* (2012) [86] introduced the *generalized beta-generated family*, which extends the BG family by substituting $G(x; \boldsymbol{\tau})$ in equation (3.1) by $G(x; \boldsymbol{\tau})^c$, where $c > 0$ is an extra shape parameter. It includes as special models the BG, Kumaraswamy-generated and exponentiated classes. Cordeiro and Brito (2012) [120] proposed the beta power distribution. They obtained explicit expressions for the moments, PWMs, mgf, mean deviations, Bonferroni and Lorenz curves, a general expression for the moments of order statistics, entropy and reliability. Cordeiro and Lemonte (2012) [121] studied the beta inverted beta distribution. The authors presented a mathematical treatment of this distribution that includes useful expansions for the density function and explicit formulae for the moments, generating and quantile functions, entropy and reliability. Cordeiro and Lemonte (2012b) [122] introduced the beta arcsine distribution, a very competitive model to the beta, beta type I and Kumaraswamy models for modelling rates and proportions. Cordeiro *et al.* (2012a) [123] studied some mathematical properties of the BW distribution. They demonstrated that the BW density is a linear mixture of Weibull densities and determined its moments and two tractable expressions for the generating function. Further, they examined the asymptotic distributions of the extreme values, and derived expressions for the mean deviations, Bonferroni and Lorenz curves, reliability and two types of entropies.

Cordeiro *et al.* (2012b) [123] proposed the beta generalized gamma distribution and Cordeiro *et al.* (2012c) [79] introduced the beta extended Weibull family, which has well-known distributions as special models, such as the generalized modified Weibull, BW, beta exponential (BE), BMW and Weibull distributions, among others.

Cordeiro *et al.* (2012d) [124] extended the Moyal distribution by defining the beta Moyal distribution and investigated some of its properties. Lemonte (2012) [125] defined the beta log-logistic distribution. Singla *et al.* (2012) [126] introduced the beta generalized Weibull distribution and Zea *et al.* (2012) [127] proposed the beta exponentiated Pareto distribution, which contains as special models the BPa and exponentiated Pareto distributions. Zea *et al.* (2012) [127] investigated some mathematical properties of their distribution like ordinary and L-moments and generating and quantile functions.

Some special BG models will be discussed in Chapter 4. For illustrative purposes, we give the cdf and pdf of five BG distributions:

- The cdf and pdf of the BE distribution are

$$F(x) = \frac{1}{B(a,b)} \int_0^{1-\exp(-\lambda x)} \omega^{a-1} (1-\omega)^{b-1} d\omega, \quad x > 0,$$

and

$$f(x) = \frac{\lambda (1 - e^{-\lambda x})^{a-1} e^{-b\lambda x}}{B(a,b)},$$

respectively, where $\lambda > 0$ is a scale parameter.

- The cdf and pdf of the BW distribution are

$$F(x) = \frac{1}{B(a,b)} \int_0^{1-\exp\{-(\lambda x)^\gamma\}} \omega^{a-1} (1-\omega)^{b-1} d\omega, \quad x > 0,$$

and

$$f(x) = \frac{\gamma \lambda^\gamma e^{-(\lambda x)^\gamma} [1 - e^{-(\lambda x)^\gamma}]^{a-1} [e^{-(\lambda x)^\gamma}]^{b-1}}{B(a,b)},$$

respectively, where $\lambda > 0$ (scale parameter) and $\gamma > 0$ (shape parameter).

- The cdf and pdf of the standard *beta Gumbel* distribution are

$$F(x) = \frac{1}{B(a,b)} \int_0^{1-\exp(-e^x)} \omega^{a-1} (1-\omega)^{b-1} d\omega, \quad x \in \mathbb{R},$$

and

$$f(x) = \frac{\exp(x - e^x)[1 - \exp(-e^x)]^{a-1} \exp(-e^x)^{b-1}}{B(a,b)},$$

respectively. We can introduce location and scale parameters to this distribution by defining $Y = \mu + \sigma X$.

- The cdf and pdf of the *beta Lomax* distribution are

$$F(x) = \frac{1}{B(a,b)} \int_0^{1-\left(\frac{\beta}{\beta+x}\right)^\alpha} \omega^{a-1} (1-\omega)^{b-1} \mathrm{d}\omega, \quad x > 0,$$

and

$$f(x) = \frac{\alpha \, \beta^\alpha (\beta + x)^{-(\alpha+1)}}{B(a,b)} \left[1 - \left(\frac{\beta}{\beta+x}\right)^\alpha\right]^{a-1} \left(\frac{\beta}{\beta+x}\right)^{(b-1)\alpha},$$

respectively, where $\alpha > 0$ and $\beta > 0$.

- The cdf and pdf of the *beta inverted beta* distribution are

$$F(x) = \frac{1}{B(a,b)} \int_0^{I_{\frac{x}{1+x}}(\alpha,\beta)} \omega^{a-1} (1-\omega)^{b-1} \mathrm{d}\omega, \quad x > 0,$$

and

$$f(x) = \frac{x^{\alpha-1} (1+x)^{-(\alpha+\beta)}}{B(\alpha,\beta) \, B(a,b)} \, I_{\frac{x}{1+x}}(\alpha,\beta)^{a-1} \left[1 - I_{\frac{x}{1+x}}(\alpha,\beta)\right]^{b-1},$$

respectively, where $\alpha > 0$ and $\beta > 0$.

3.2. SOME SPECIAL MODELS

In Table 3.2, we present some BG density functions by taking the following baseline models: normal, log-normal, Student-t, Laplace, exponential, Weibull, Gumbel, Birnbaum Saundors, gamma, Pareto and logistic distributions. More recently, several papers on BG distributions were published. Barros *et al.* (2013) [128] proposed the beta generalized distributions and related exponentiated models. Cordeiro *et al.* (2013a) [129] proposed the beta Weibull geometric (BWG) distribution, which has important properties written in terms of the Weibull properties since the BWG density is a mixture of Weibull densities. Expansions for the moments, generating function, mean deviations and Rényi entropy were provided by the authors. Cordeiro *et al.* (2013b) [130] introduced and studied the beta generalized Rayleigh (BGR) distribution that contains as special cases some important models. The authors obtained expanded quantities for the cumulative distribution and density functions. Cordeiro *et al.* (2013c) [131] introduced the beta exponentiated Weibull distribution. This distribution is an important competitive model to the Weibull, exponentiated Weibull, beta exponential and BW distributions.

Gomes *et al.* (2013) [132] considered the beta Burr III (BBIII) distribution. They demonstrated that the BBIII pdf can be expressed as a linear mixture of Burr III densities and then some of its structural properties are easily obtained. Cordeiro *et al.* (2014) [133] proposed the beta extended half-normal model, which extends the exponential, extended half-normal and half-normal distributions. Lemonte and Cordeiro (2013) [134] introduced the beta Lomax distribution and derived some of its structural properties, such as explicit expressions for the moments, generating and quantile functions. Finally, Shittu and Adepoju (2013) [135] combined the Nakagami and beta distributions in order to define a model that is better than each of them individually in terms of the estimation of their characteristics and parsimonious in their parameters based on the BG generator. This model is called the beta-Nagakami distribution.

3.3. QUANTILE FUNCTION

The qfs are very useful in Statistics to obtain some mathematical and statistical properties of continuous distributions. The importance of these functions is well described in the book by Gilchrist (2000) [136].

The qf of the BG family, say $Q(u) = Q(u; a, b, \boldsymbol{\tau})$, can be determined in general by inverting the baseline G cumulative distribution. Then, it can be expressed in terms of the baseline qf as

$$Q(u) = Q_G\left(Q_\beta(u)\right), \quad u \in (0, 1), \tag{3.3}$$

where $Q_G(u) = G^{-1}(u)$ is the parent qf and $Q_\beta(u) = I^{-1}(u; a, b)$ is the beta qf with positive parameters a and b. It is possible to obtain a power series for $Q_\beta(u)$. One of them can be found on the Wolfram website at `http://functions.wolfram.com/06.23.06.0004.01` as

$$z = Q_\beta(u) = a_1 v + a_2 v^2 + a_3 v^3 + a_4 v^4 + O(v^{5/a}),$$

where $v = [a\, B(a, b)\, u]^{1/a}$ (for $a > 0$), $a_0 = 0$, $a_1 = 1$, $a_2 = (b-1)/(a+1)$,

$$a_3 = (b-1)[a^2 + (3b-1)a + 5b - 4]/[2(a+1)^2(a+2)],$$

$$a_4 = (b-1)[a^4 + (6b-1)a^3 + (b+2)(8b-5)a^2 + (33b^2 - 30b + 4)a \\ + b(31b - 47) + 18]/[3(a+1)^3(a+2)(a+3)], \dots.$$

The coefficients a_i's (for $i \geq 2$) can be derived from a cubic recursion of the form

$$a_i =$$

$$\frac{1}{[i^2 + (a-2)i - (a-1)]} \left\{ (1 - \delta_{i,2}) \sum_{r=2}^{i-1} a_r \, a_{i+1-r} \left[r(1-a)(i-r) - r(r-1) \right] \right.$$

$$\left. + \sum_{r=1}^{i-1} \sum_{s=1}^{i-r} a_r \, a_s \, a_{i+1-r-s} \left[r(r-a) + s(a+b-2)(i+1-r-s) \right] \right\},$$

where $\delta_{i,2} = 1$ (if $i = 2$) and $\delta_{i,2} = 0$ (if $i > 2$). In the last equation, we note that the quadratic term only contributes for $i \geq 3$.

Note that the qf $Q(u)$ can be used to measure the effects of the shape parameters a and b on the skewness and kurtosis of the random variable X. Kenney and Keeping (1962) [137] suggested the Bowley skewness, one of the first skewness measures, which is defined by the average of the quartiles minus the median divided by half the interquartile range:

$$B = \frac{Q(3/4) + Q(1/4) - 2Q(1/2)}{Q(3/4) - Q(1/4)}.$$

This formula is more tractable than the skewness original definition. The Moors' kurtosis is based on octiles

$$M = \frac{Q(7/8) - Q(5/8) + Q(3/8) - Q(1/8)}{Q(6/8) - Q(2/8)}.$$

The above statistics have lower sensitivity to outliers and their existence is guaranteed even for distributions without moments. The measure M is not sensitive to variations of the values in the tails or to variations of the values around the median. For fixed $Q(2/8)$ and $Q(6/8)$, M decreases as $Q(3/8) - Q(1/8)$ and $Q(7/8) - Q(5/8)$ decrease. So, if $Q(3/8) \to Q(1/8)$ and $Q(7/8) \to Q(5/8)$, then M goes to zero and half of the total probability mass is concentrated in the neighborhoods of the octiles $Q(2/8)$ and $Q(6/8)$. Then, $M > 0$ and there is a good agreement with the usual kurtosis measures for some distributions. For the normal distribution, $B = M = 0$.

Table 3.2: Special BG distributions. In the Student-t distribution, $\lambda = a + b$.

Distribution	Baseline Distribution	BG Density
Normal	$G(x) = \Phi(x)$	$f(x) = \dfrac{\phi(x)\Phi(x)^{a-1}[1-\Phi(x)]^{b-1}}{B(a,b)}$
Log-Normal	$G(x) = \Phi(\log x)$	$f(x) = \dfrac{\phi(\log x)\Phi(\log x)^{a-1}[1-\Phi(\log x)]^{b-1}}{x\,B(a,b)}$
Student t	$G(x) = \dfrac{1}{2}\left(1+\dfrac{x}{\sqrt{\lambda+x^2}}\right)$	$f(x) = \dfrac{\lambda}{2^a B(a,b)}\dfrac{1}{(\lambda+x^2)^{3/2}}\left(1+\dfrac{x}{\sqrt{\lambda+x^2}}\right)^{a-1}\left[1-\dfrac{1}{2}\left(1+\dfrac{x}{\sqrt{\lambda+x^2}}\right)\right]^{b-1}$
Laplace	$G(x) = \begin{cases} \frac{1}{2}\exp(x/\lambda), & x<0, \\ 1-\frac{1}{2}\exp(-x/\lambda), & x>0 \end{cases}$	$f(x) = \dfrac{g(x)G(x)^{a-1}[1-G(x)]^{b-1}}{B(a,b)}$
Exponential	$G(x) = 1-\exp(-\lambda x),\,\lambda>0$	$f(x) = \dfrac{\lambda e^{-\lambda x}[1-e^{-\lambda x}]^{a-1}[1-(1-e^{-\lambda x})]^{b-1}}{B(a,b)}$
Weibull	$G(x) = 1-\exp[-(\lambda x)^\gamma],\,\gamma>0$	$f(x) = \dfrac{\gamma\lambda^\gamma e^{-(\lambda x)^\gamma}[1-e^{-(\lambda x)^\gamma}]^{a-1}[1-(1-e^{-(\lambda x)^\gamma})]^{b-1}}{B(a,b)}$
Gumbel	$G(x) = 1-\exp(-e^x)$	$f(x) = \dfrac{\exp(x-e^x)[1-\exp(-e^x)]^{a-1}[1-(1-\exp(-e^x))]^{b-1}}{B(a,b)}$

Distribution	Baseline Distribution	BG Density
Birnbaum-Saunders	$G(x) = \Phi\left[\frac{1}{\alpha}\left(\sqrt{\frac{x}{\beta}} - \sqrt{\frac{\beta}{x}}\right)\right]$, $\alpha, \beta > 0$	$f(x) = \frac{g(x)[G(x)]^{a-1}[1-G(x)]^{b-1}}{B(a,b)}$
Gamma	$G(x) = \gamma_1(\alpha, \beta x) = \frac{\gamma(\alpha,\beta x)}{\Gamma(\alpha)}$, $\alpha, \beta > 0$	$f(x) = \frac{\beta^\alpha x^{\alpha-1} e^{-\beta x} \gamma_1(\alpha,\beta x)^{a-1}[1-\gamma_1(\alpha,\beta x)]^{b-1}}{B(a,b)\Gamma(\alpha)}$
Pareto	$G(x) = 1 - \frac{1}{(1+x)^\nu}$, $\nu > 0$	$f(x) = \nu\frac{\left[1-\frac{1}{(1+x)^\nu}\right]^{a-1}\left[1-\left(1-\frac{1}{(1+x)^\nu}\right)\right]^{b-1}}{(1+x)^{\nu+1}B(a,b)}$
Logistic	$G(x) = \frac{1}{1+e^{-x}}$	$f(x) = \frac{e^{-x}\left[1-(1+e^{-x})^{-1}\right]^{b-1}}{(1+e^{-x})^{a+1}B(a,b)}$

3.4. USEFUL EXPANSIONS

In this section, we present a useful expansion for the BG density function. Note that the generalized binomial coefficient to real arguments is defined by $\binom{a}{b} = \Gamma(a+1)/[\Gamma(b+1)\Gamma(a-b+1)]$. Expanding the binomial in (3.2) gives

$$f(x) = B(a,b)^{-1} g(x) \sum_{i=0}^{\infty} (-1)^i \binom{b-1}{i} G(x)^{a+i-1}. \tag{3.4}$$

Hence, the BG cumulative function can be expressed as a mixture of exp-G cdfs given by

$$F(x) = \sum_{i=0}^{\infty} w_i \, H_{a+i}(x), \tag{3.5}$$

whereas the density function is simply

$$f(x) = \sum_{i=0}^{\infty} w_i \, h_{a+i}(x), \tag{3.6}$$

where

$$w_i = \frac{(-1)^i \binom{b-1}{i}}{(a+i) \, B(a,b)}.$$

Here, $h_{(a+i)}(x)$ and $H_{(a+i)}(x)$ denote the pdf and cdf of the exp-G distribution with power parameter $(a+i)$. Based on the linear combinations (3.5) and (3.6), several properties (like ordinary and incomplete moments, generating function, mean deviations, *etc.*) of the BG family can be obtained directly from those properties of the exp-G distribution.

The properties of some exp-G distributions have been studied by several authors. We refer the reader to Mudholkar and Srivastava (1993) [7], Mudholkar *et al.* (1995) [36], Gupta *et al.* (1998) [37], Gupta and Kundu (1999) [42], Gupta and Kundu (2001) [43], Nadarajah and Gupta (2007) [39], Gupta and Gupta (2008) [138], Carrasco *et al.* (2008) [5], Barreto–Souza and Cribari–Neto (2009) [33], Silva *et al.* (2010a) [9], AL-Hussaini (2010) [139], Gusmão *et al.* (2011) [140], Lemonte and Cordeiro (2011) [141], Cordeiro *et al.* (2011a) [40], Al-Hussaini and Hussain (2011) [142], Lemonte *et al.* (2013) [143] and Lemonte (2013) [144], among several others.

3.5. MOMENTS

Let $X \sim \mathrm{BG}(a,b,\boldsymbol{\tau})$ and $\mu'_s = \mathrm{E}(X^s)$ be its sth moment. We provide three general formulae for the ordinary moments of X. First, it can be obtained as

$\mu'_s = E[\{Q_G(U)\}^s]$, where $U \sim \text{Beta}(a, b)$. Then, we have

$$\mu'_s = B(a, b)^{-1} \int_0^1 [Q_G(u)]^s \, u^{a-1} \, (1 - u)^{b-1} du. \tag{3.7}$$

Let Y have the baseline G distribution. An approximation to μ'_s follows by expanding $Q_G(u)$ in Taylor series around the point $E(Y) = \mu_G$:

$$\mu'_s \approx \sum_{k=0}^{s} \binom{s}{k} [Q_G(\mu_G)]^{s-k} [Q_G^{(1)}(\mu_G)]^k \sum_{i=0}^{k} (-1)^i \binom{k}{i} \mu_G^i \, \frac{B(a + k - i, b)}{B(a, b)},$$

where $Q_G^{(1)}(u) = dQ_G(u)/du = [g(Q_G(u))]^{-1}$.

A second alternative expression for μ'_s can be determined in terms of the PWMs of Y, say $\tau_{s,k} = E[Y^s \, G(Y)^k]$, for $s, k = 0, 1, \ldots$ We can write

$$\tau_{s,k} = \int_{-\infty}^{\infty} x^s \, G(x)^k \, g(x) dx = \int_0^1 [Q_G(u)]^s \, u^k du.$$

The quantities $\tau_{s,k}$ can be evaluated at least numerically for most baseline distributions.

For $a > 0$ integer, it follows from equation (3.6) that

$$\mu'_s = B(a, b + 1)^{-1} \sum_{i=0}^{\infty} (-1)^i \binom{b}{i} \tau_{s, a+i-1}.$$

Otherwise, we can expand $F(x)^\alpha$ (for real $\alpha > 0$) as

$$F(x)^\alpha = \sum_{k=0}^{\infty} t_k \, F(x)^k, \tag{3.8}$$

where

$$t_k = t_k(\alpha) = \sum_{j=k}^{\infty} (-1)^{k+j} \binom{\alpha}{j} \binom{j}{k}.$$

Then, using (3.8) in equation (3.6) gives

$$\mu'_s = \sum_{k=0}^{\infty} v_k \, \tau_{s,k}, \tag{3.9}$$

where

$$v_k = v_k(a, b) = B(a, b + 1)^{-1} \sum_{i=0}^{\infty} \sum_{j=k}^{\infty} (-1)^{k+i+j} \binom{a + i - 1}{j} \binom{j}{k} \binom{b}{i}.$$

Next, we provide two applications of equation (3.9). The moments of the BE distribution (with parameter $\lambda > 0$) are

$$\mu'_s = s!\,\lambda^s \sum_{k,j=0}^{\infty} \frac{(-1)^{s+j}\,v_k}{(j+1)^{s+1}} \binom{k}{j}.$$

For the beta standard logistic (BSL) distribution, where the baseline cumulative distribution is $G(x) = (1 + e^{-x})^{-1}$, based on a result by Prudnikov *et al.* (1986) [145], we obtain (for $t < 1$)

$$\mu'_s = \sum_{k=0}^{\infty} v_k \left.\frac{\partial^s}{\partial t^s} B(t + k + 1, 1 - t)\right|_{t=0}.$$

For example, we have

$$\left.\frac{\partial}{\partial t} B(t + k + 1, 1 - t)\right|_{t=0} = \frac{\gamma + \psi(1 + k)}{1 + k},$$

$$\left.\frac{\partial^2}{\partial t^2} B(t + k + 1, 1 - t)\right|_{t=0} = \frac{\pi^2/6 + \psi'(1 + k)}{1 + k} + \frac{[\gamma + \psi(1 + k)]^2}{1 + k},$$

where $\psi(\cdot)$ is the digamma function, $\psi'(\cdot)$ is the trigamma function and γ is the Euler's constant.

Finally, a third representation for μ'_s (for $a > 0$ integer) can be derived from (3.6) as

$$\mu'_s = B(a, b + 1)^{-1} \sum_{i=0}^{\infty} (-1)^i \binom{b}{i} \tau_{s,i+a-1}. \qquad (3.10)$$

Equations (3.7), (3.9) and (3.10) are the main results of this section. The central moments, cumulants, and other types of moments can be obtained from the general formulae given in Section 1.5.

The sth incomplete moment of X is defined by

$$m_s(y) = \int_{-\infty}^{y} x^s\, F(x)^k\, f(x)\mathrm{d}x,$$

which can be determined by combining (3.6) and (3.9). We have

$$m_s(y) = \sum_{k=0}^{\infty} v_k \int_0^{G(y)} Q_G(u)^s\, u^k\, du,$$

where this integral can be computed for most baseline distributions.

3.6.　SOME BASELINE PWMs

For several BG distributions, the quantities $\tau_{s,r}$ are simple to be obtained as infinite sums of well-known special functions. In the next sections, as illustration, we present explicit expressions for the PWMs of the beta gamma, beta normal, beta beta and beta Student-t distributions. We choose these distributions since they are the most popular and they guarantee the existence of the moments of the generated BG distributions.

Some special BG pdfs are described in Section 3.2. Similar results could be derived for other of these distributions. These closed-form expressions may be employed to deduce other linear representations for the mgf, cgf, cf, factorial and central moments.

Several baseline PWMs involve special functions including well-known functions defined in Chapter 1.

3.6.1.　PWMs of the Beta Gamma

Here, we obtain $\tau_{s,r}$ using a power series for the incomplete gamma function. After some algebra, we obtain

$$\tau_{s,r} = \frac{\beta^{-s}}{\Gamma(\alpha)^{r+1}} \int_0^\infty u^{s+\alpha-1} \, e^{-u} \left[u^\alpha \sum_{m=0}^\infty \frac{(-u)^m}{(\alpha+m)m!} \right]^r du.$$

The last integral can be obtained from equations (24) and (25) of Nadarajah (2008) [146] as

$$\tau_{s,r} = \frac{\beta^{-s}\alpha^{-r}\Gamma(s+\alpha(r+1))}{\Gamma(\alpha)^{r+1}}$$
$$\times F_A^{(r)}(s+\alpha(r+1); \alpha, \ldots, \alpha; \alpha+1, \ldots, \alpha+1; -1, \ldots, -1).$$

Hence, the moments of the beta gamma distribution can be expressed as infinite weighted sums of the Lauricella functions of type A (see (1.10) in Section 1.4).

3.6.2.　PWMs of the Beta Normal

We work with the standard BN in generality, since we can obtain the moments of $X \sim \mathrm{BN}(a, b, \mu, \sigma)$ from the moments of $Z \sim \mathrm{BN}(a, b, 0, 1)$ using $\mathrm{E}(X^r) =$

$\mathrm{E}[(\mu + \upsilon Z)^r] \to \sum_{t=0}^{r} \mu^{r-t} \sigma^r \mathrm{F}(Z^r)$. For s and r positive integers, we define

$$\tau_{s,r} = \int_{-\infty}^{\infty} x^s \phi(x) \Phi(x)^r \, dx.$$

Using the binomial expansion and interchanging terms, we have

$$\tau_{s,r} = \frac{1}{2^r \sqrt{2\pi}} \sum_{l=0}^{r} \binom{r}{l} \int_{-\infty}^{\infty} x^s e^{-x^2/2} \operatorname{erf}\left(\frac{x}{\sqrt{2}}\right)^{r-l} dx.$$

Based on the power series for the error function

$$\operatorname{erf}(x) = \frac{2}{\sqrt{\pi}} \sum_{m=0}^{\infty} \frac{(-1)^m x^{2m+1}}{(2m+1)m!},$$

we determine the last integral following equations (9)-(11) of Nadarajah (2008). We can obtain when $s + r - l$ is even

$$\tau_{s,r} = 2^{s/2} \pi^{-(r+1/2)} \sum_{\substack{l=0 \\ (s+r-l)\,\text{even}}}^{r} 2^{-l} \pi^l \binom{r}{l} \Gamma\left(\frac{s+r-l+1}{2}\right) \times$$

$$F_A^{(r-l)}\left(\frac{s+r-l+1}{2}; \frac{1}{2}, \ldots, \frac{1}{2}; \frac{3}{2}, \ldots, \frac{3}{2}; -1, \ldots, -1\right).$$

The terms in $\tau_{s,r}$ vanish when $s + r - l$ is odd.

3.6.3. PWMs of the Beta Beta

The (s,r)th PWM of the Beta(α, β) distribution is given by

$$\tau_{s,r} = \frac{1}{B(\alpha, \beta)} \int_0^1 x^{s+\alpha-1} (1-x)^{\beta-1} I_x(\alpha, \beta)^r \, dx.$$

Using the power series for the incomplete beta function for β real non-integer

$$I_x(\alpha, \beta) = \frac{x^\alpha}{B(\alpha, \beta)} \sum_{m=0}^{\infty} \frac{(1-\beta)_m x^m}{(\alpha+m)m!},$$

and $(f)_k = \Gamma(f+k)/\Gamma(f)$, we can evaluate the above integral following similar approach by Nadarajah (2008, Section 5) [146]. In fact, the function $I(k,l)$ defined in equation (28) can be given in terms of the generalized Kampé de Fériet function (see Section 1.11).

Hence, we obtain

$$
\begin{aligned}
\tau_{s,r} &= \alpha^{-r} B(\alpha,\beta)^{-(r+1)} B(\beta, s + \alpha(r+1)) \times \\
&\quad F_{1:1}^{1:2}\Big((s + \alpha(r+1)) : (1 - \beta, \alpha); \ldots ; (1 - \beta, \alpha) : \\
&\quad (\beta + s + \alpha(r+1)) : (\alpha + 1); \ldots ; (\alpha + 1); 1, \ldots, 1\Big).
\end{aligned}
$$

The moments of the beta beta distribution follow immediately as infinite sums of the generalized Kampé de Fériet functions.

3.6.4.   PWMs of the Beta Student t

For any real x, the cdf of the Student t distribution is simply $G(x) = I_y(1/2, \nu/2)$, where $y = (x + \sqrt{x^2 + \nu})/(2\sqrt{x^2 + \nu})$ and $\nu > 0$ denotes the degrees of freedom. Let $g(x) = \mathrm{d}G(x)/\mathrm{d}x$ be the Student t pdf. Since the Student t density is symmetric around zero, its (s,r)th PWM can be expressed as

$$
\tau_{s,r} = \int_0^\infty x^s\, G(x)^r\, g(x) dx + (-1)^s \int_0^\infty x^s\, [1 - G(x)]^r\, g(x) dx,
$$

For k, n and m positive integers, we define

$$
A_{k,n,m} = \int_0^\infty x^k\, G(x)^{m-1}\, [1 - G(x)]^{n-m}\, g(x) dx
$$

and then rewrite $\tau_{s,r}$ as

$$
\tau_{s,r} = A_{s,r+1,r+1} + (-1)^s\, A_{s,r+1,1}.
$$

For $x \geq 0$, we have $G(x) = \frac{1}{2} + \frac{1}{2} I_{x^2/(\nu+x^2)}(1/2, \nu/2)$. By setting $y = x^2/(\nu+x^2)$ and using the incomplete beta function expansion, we can obtain the integral $A_{k,n,m}$ in terms of the generalized Kampé de Fériet function; see Nadarajah (2007) [147]. Then, from equation (7) of his paper, we obtain $A_{s,r+1,r+1}$ and $A_{s,r+1,1}$. By combining these expressions, we can write

$$
\begin{aligned}
\tau_{s,r} &= \frac{\nu^{s/2}}{2^{r+1}} \sum_{\substack{p=0 \\ p \text{ even}}}^{r} \binom{r}{p} 2^{p+1} B^{-1-p}(1/2, \nu/2) B\left(\frac{\nu - s}{2}, \frac{s + p + 1}{2}\right) \times \\
&\quad F_{1:1}^{1:2}\Bigg(\left(\frac{s+p+1}{2}\right) : \left(1 - \frac{\nu}{2}, \frac{1}{2}\right); \ldots ; \left(1 - \frac{\nu}{2}, \frac{1}{2}\right); \\
&\quad \left(\frac{\nu + p + 1}{2}\right) : \left(\frac{3}{2}\right); \ldots ; \left(\frac{3}{2}\right); 1, \ldots, 1\Bigg).
\end{aligned}
$$

Hence, the moments of the beta Student t distribution can be expressed as infinite weighted sums of the generalized Kampé de Fériet functions (see (1.11) in Section 1.4).

3.7. PWMs BASED ON QUANTILES

For $b > 0$ real non-integer, we have the power series

$$(1 - u)^{b-1} = \sum_{j=0}^{\infty} (-1)^j \binom{b-1}{j} u^j,$$

where the binomial coefficient is defined for any real. The nth moment of X is given in terms of the baseline qf $Q_G(u)$ as

$$E(X^n) = \frac{1}{B(a,b)} \int_0^1 Q_G(u)^n \, u^{a-1} (1-u)^{b-1} du.$$

Thus,

$$E(X^n) = \frac{1}{B(a,b)} \sum_{j=0}^{\infty} (-1)^j \binom{b-1}{j} \int_0^1 Q_G(u)^n \, u^{a+j-1} du. \qquad (3.11)$$

If b is a positive integer, the index j in the above sum stops at $b-1$. Equation (3.11) can be applied to several others BG distributions. For some baseline distributions with closed-forms for $Q_G(u)$, we can evaluate the integral in (3.11). However, for some other distributions, the solution is not possible.

Next, we provide two simple applications of (3.11). First, we consider the BW distribution (with four positive parameters) defined by taking the cumulative function $G_{\lambda,c}(x) = 1 - \exp\{-(\lambda x)^c\}$ of the Weibull distribution with shape parameter $c > 0$ and scale parameter $\lambda > 0$. The BW density function has the form

$$f(x) = \frac{c\lambda^c}{B(a,b)} \, x^{c-1} \exp\{-b(\lambda x)^c\} \, [1 - \exp\{-(\lambda x)^c\}]^{a-1}.$$

The corresponding qf is $Q_G(u) = \lambda^{-1} \left[-\log(1-u)\right]^{1/c}$ and then

$$E(X^n) = \frac{1}{\lambda^{n/c} B(a,b)} \sum_{j=0}^{\infty} (-1)^j \binom{b-1}{j} \int_0^1 (1-z)^{a+j-1} \log^{n/c}(1/z) dz.$$

By expanding the binomial term, we have

$$E(X^n) = \frac{1}{\lambda^{n/c} B(a,b)} \sum_{j,k=0}^{\infty} (-1)^{j+k} \binom{b-1}{j} \binom{a+j-1}{k}$$

$$\times \int_0^1 z^k \log^{n/c}(1/z) dz.$$

Using equation (2.6.3.1) in Prudnikov *et al.* (1986) [145], we obtain

$$\mathrm{E}(X^n) = \frac{1}{\lambda^{n/c}\, B(a,b)} \sum_{j,k=0}^{\infty} \frac{(-1)^{j+k}\, \Gamma(n/c+1)}{(k+1)^{n/c+1}} \binom{b-1}{j} \binom{a+j-1}{k}.$$

As a second example, we consider the BSL distribution for which the cumulative and quantile functions are $G(x) = (1 + \mathrm{e}^{-x})^{-1}$ and $Q_G(u) = \log\{u/(1-u)\}$, respectively. Thus,

$$\mathrm{E}(X^n) = \frac{1}{B(a,b)} \sum_{j=0}^{\infty} (-1)^j \binom{b-1}{j} \int_0^1 z^{a+j-1} \log^n\{z/(1-z)\}dz.$$

The above integral can be determined numerically using `Maple`.

Power series methods might have great importance in Applied Mathematics and Statistics. We provide a new method for the moments of the BG family based on power series for the parent qf $x = Q_G(u)$, whose coefficients can be computed by nonlinear recurrence equations.

If the function $Q_G(u)$ has a closed-form expression, it can be given as a power series of a transformed variable $v = p(qu - t)^\rho$, where p, q, t and ρ are known constants. We can write

$$Q_G(u) = \sum_{i=0}^{\infty} a_i\, v^i, \tag{3.12}$$

where the a_i's are real numbers. For the normal, beta, gamma and Student t distributions, among others, $Q_G(u)$ has not simple expressions but all of them can be expanded as in equation (3.12).

By application of an equation in Section 0.314 of Gradshteyn and Ryzhik (2000) [11] for a power series raised to any positive integer power n, we have

$$Q(u)^n = \left(\sum_{i=0}^{\infty} a_i\, v^i \right)^n = \sum_{i=0}^{\infty} c_{n,i}\, v^i. \tag{3.13}$$

Here, the coefficients $c_{n,i}$, for $i = 1, 2, \ldots$, are easily determined from the recurrence equation

$$c_{n,i} = (i\, a_0)^{-1} \sum_{m=1}^{i} [m(n+1) - i]\, a_m\, c_{n,i-m}, \tag{3.14}$$

and $c_{n,0} = a_0^n$. The coefficient $c_{n,i}$ follows from $c_{n,0}, \ldots, c_{n,i-1}$ and hence from the quantities $a_0, \ldots, a_i$. Equations (3.13) and (3.14) are used throughout this book. The coefficient $c_{n,i}$ can be given clearly in terms of the coefficients a_i's. However, numerical programming in algebraic or numerical software make such expressions unnecessary.

From equations (3.11) and (3.13), we obtain

$$\mathrm{E}(X^n) = \frac{1}{B(a,b)} \sum_{i,j=0}^{\infty} (-1)^j \binom{b-1}{j} c_{n,i}\, I_{i,j}, \qquad (3.15)$$

where the integral has a simple form $I_{i,j} = \int_0^1 v^i\, u^{a+j-1} du$ and the quantities $c_{n,i}$ are easily obtained from the recurrence equation (3.14).

Established algebraic expansions to determine the moments of any BG distribution can be more efficient than computing them directly by numerical integration of the expression

$$\mu_n' = \frac{1}{B(a,b)} \int_{-\infty}^{\infty} x^n\, g(x)\, G(x)^{a-1} \{1 - G(x)\}^{b-1} dx,$$

which can be prone to rounding off errors among others.

We provide below three applications of equation (3.15).

3.7.1. Moments of the Beta Gamma

The cdf of a random variable Y with a gamma distribution with shape parameter $\alpha > 0$ and scale parameter $\beta > 0$ is

$$G_{\alpha,\beta}(x) = \frac{\gamma(\alpha, \beta x)}{\Gamma(\alpha)}, \qquad x > 0.$$

We have $E(Y^n) = \Gamma(n+\alpha)/\Gamma(\beta^n \alpha)$. A random variable X with a beta gamma distribution, denoted as $\mathrm{BG}(a,b,\alpha,\beta)$, has density function

$$f(x) = \frac{\beta^\alpha x^{\alpha-1}\, e^{-\beta x}}{B(a,b)\Gamma(\alpha)^{a+b-1}}\, \gamma(\alpha, \beta x)^{a-1} \{\Gamma(\alpha) - \gamma(\alpha, \beta x)\}^{b-1}, \qquad x > 0.$$

If W has a gamma $G_{\alpha,1}(x)$ distribution, then $Y = W/\beta$ follows a gamma $G(\alpha, \beta)$ distribution. The qf of Y, say $Q_Y(v)$, can be determined from the qf of W, say $Q_W(v)$, by $Q_Y(v) = Q_W(\beta v)$ and it is given by (3.12) with

$v = \beta[u\,\Gamma(\alpha+1)]^{1/\alpha}$, $a_0 = 0$, $a_1 = 1$ and any coefficient a_{i+1} (for $i \geq 1$) is determined by the cubic recurrence equation

$$
a_{i+1} = \frac{1}{i(\alpha+i)}\left\{\sum_{r=1}^{i}\sum_{s=1}^{i-s+1} a_r\,a_s\,a_{i-r-s+2}\,s\,(i-r-s+2)\right.
$$
$$
\left. - \Delta(i)\sum_{r=2}^{i} a_r\,a_{i-r+2}\,r\,[r-\alpha-(1-\alpha)(i+2-r)]\right\},
$$

where $\Delta(i) = 0$ if $i < 2$ and $\Delta(i) = 1$ if $i \geq 2$. The first few coefficients are $a_2 = 1/(\alpha+1)$, $a_3 = (3\alpha+5)/[2(\alpha+1)^2(\alpha+2)],\ldots$ The quantities $c_{n,i}$'s can be evaluated numerically from these a_i's and the integral $I_{i,j}$ in (3.15) becomes

$$
I_{i,j} = \frac{\Gamma(\alpha+1)^{i/\alpha}}{a+i/\alpha+j}.
$$

Hence, the moments of the BG(a, b, α, β) distribution are

$$
E(X^n) = \frac{1}{\beta^n B(a,b)}\sum_{i,j=0}^{\infty}\frac{(-1)^j\,c_{n,i}\,\Gamma(\alpha+1)^{i/\alpha}}{(a+i/\alpha+j)}\binom{b-1}{j}.
$$

3.7.2.　Moments of the Beta Student t

The Student t (St) distribution is one of the most common continuous models in Statistics and applied areas. The density function of the Student t_ν distribution with $\nu > 0$ degrees of freedom is given by

$$
g(x) = \frac{1}{\sqrt{\nu}B(1/2,\nu/2)}\left(1+\frac{x^2}{\nu}\right)^{-(\nu+1)/2},
$$

for $x \in \mathbb{R}$. The cdf of the Student t_ν distribution is simply $G(x) = I_y(1/2,\nu/2)$, where $y = (x+\sqrt{x^2+\nu})/(2\sqrt{x^2+\nu})$. The beta Student BSt$(a,b,\nu)$ density (for $x \in \mathbb{R}$) with parameters a, b and ν is

$$
f(x) = \frac{\left(1+\frac{x^2}{\nu}\right)^{-(\nu+1)/2}}{\sqrt{\nu}B(a,b)B(1/2,\nu/2)}\,I_{\left(\frac{x+\sqrt{x^2+\nu}}{2\sqrt{x^2+\nu}}\right)}(1/2,\nu/2)^{a-1}
$$
$$
\times\left\{1-I_{\left(\frac{x+\sqrt{x^2+\nu}}{2\sqrt{x^2+\nu}}\right)}(1/2,\nu/2)\right\}^{b-1}.
$$

The beta Student t is symmetric around zero only when $a = b$.

The qf of the BSt distribution also follows (3.12) with

$$v = \sqrt{\nu\pi}\,\frac{\Gamma(\nu)}{\Gamma(\frac{\nu+1}{2})}(u - 1/2).$$

Here, $b_0 = 1$ and the coefficients b_k's (for $k \geq 1$) are given by a cubic recurrence equation (Shaw, 2006) [148]

$$
\begin{aligned}
b_k &= \frac{1}{2k(2k+1)}\sum_{r=0}^{k-1}\sum_{s=0}^{k-r-1} b_r\,b_s\,b_{k-r-s-1} \\
&\times \left\{(1+\nu^{-1})[(2s+1)(2k-2r-2s-1)] - 2r(2r+1)\nu^{-1}\right\}.
\end{aligned}
$$

It is readily verified that the first coefficients are: $b_1 = (\nu+1)/(6\nu)$, $b_2 = (7\nu^2 + 8\nu + 1)/(120\nu^2)$, $b_3 = (127\nu^3 + 135\nu^2 + 9\nu + 1)/(5040\nu^3)\ldots$ The coefficients a_i's can be obtained from these b_i's and then the quantities $c_{n,i}$'s can be determined numerically from the a_i's. The integral $I_{i,j}$ becomes

$$I_{i,j} = (\nu\pi)^{i/2}\frac{\Gamma(\nu)^i}{\Gamma(\frac{\nu+1}{2})^i}\sum_{k=0}^{i}\frac{(-1)^k\,2^{-k}}{(a+i+j-k)}\binom{i}{k}.$$

The moments of the BSt model follow from the above results.

3.7.3. Moments of the Beta Beta

The beta distribution is the most important distribution for modeling proportions. It has applications in several areas such as hydrology and economics. There are many generalizations of the beta distribution and some of them involve algebraic, exponential and hypergeometric functions. The reader is refereed to Gupta and Nadarajah's (2004) book [102].

The pdf and cdf of the beta $B(\alpha, \beta)$ distribution with parameters $\alpha > 0$ and $\beta > 0$ are $g(x) = x^{\alpha-1}(1-x)^{\beta-1}/B(\alpha, \beta)$ and $G(x) = I_x(\alpha, \beta)$ for $x \in (0, 1)$, respectively. In this section, an enlargement of the beta distribution is presented called the beta beta $BB(a, b, \alpha, \beta)$ distribution. It has density function given by

$$f(x) = \frac{x^{\alpha-1}(1-x)^{\beta-1}}{B(\alpha, \beta)\,B(a, b)}I_x(\alpha, \beta)^{a-1}\left\{1 - I_x(\alpha, \beta)\right\}^{b-1}, \quad x \in (0, 1).$$

One of the expansions for the inverse of the incomplete beta function can be found in Wolfram website [1] as

$$I_x^{-1}(\alpha, \beta) = a_1 v + a_2 v^2 + a_3 v^3 + a_4 v^4 + O(v^{5/\alpha}),$$

---

[1]http://functions.wolfram.com/06.23.06.0004.01

where $v = [\alpha u B(\alpha, \beta)]^{1/\alpha}$ for $\alpha > 0$ and $a_0 = 0$, $a_1 = 1$, $a_2 = (\beta - 1)/(\alpha + 1)$,

$$a_3 = \frac{(\beta - 1)(\alpha^2 + 3\beta\alpha - \alpha + 5\beta - 4)}{2(\alpha + 1)^2(\alpha + 2)},$$

$$\begin{aligned} a_4 &= (\beta - 1)[\alpha^4 + (6\beta - 1)\alpha^3 + (\beta + 2)(8\beta - 5)\alpha^2 + (33\beta^2 - 30\beta + 4)\alpha \\ &+ \beta(31\beta - 47) + 18]/[3(\alpha + 1)^3(\alpha + 2)(\alpha + 3)], \ldots \end{aligned}$$

The coefficients a_i's for $i \geq 2$ can be derived from a cubic recursion of the form

$$\begin{aligned} a_i &= \frac{1}{[i^2 + (\alpha - 2)i + (1 - \alpha)]} \left\{ (1 - \delta_{i,2}) \sum_{r=2}^{i-1} a_r\, a_{i+1-r}\, [r(1 - \alpha)(i - r) \right. \\ &- r(r - 1)] + \sum_{r=1}^{i-1}\sum_{s=1}^{i-r} a_r\, a_s\, a_{i+1-r-s}\, [r(r - \alpha) + s(\alpha + \beta - 2) \\ &\times \left. (i + 1 - r - s)] \right\}, \end{aligned}$$

where $\delta_{i,2} = 1$ if $i = 2$ and $\delta_{i,2} = 0$ if $i > 2$. In the above equation, we note that the quadratic term contributes only for $i \geq 3$.

3.8.   GENERATING FUNCTION

In this section, we provide two formulae for the mgf of X, say $M(t) = E(e^{tX})$. A first representation for $M_X(t)$ is obtained from equation (3.6) as an infinite weighted sum

$$M_X(t) = \sum_{k=0}^{\infty} w_k\, M_{a+i}(t),$$

where $M_{a+i}(t)$ denotes the generating function of the random variable Y_{a+i} having the exp-G density function $h_{a+i}(x)$.

A second formula for $M(t)$ follows from (3.6) as

$$M(t) = \sum_{k=0}^{\infty}(a + k)\, w_k\, \rho(t, k + a - 1), \tag{3.16}$$

where $\rho(t, a)$ can be determined from the baseline $Q_G(u)$ as

$$\rho(t, a) = \int_0^1 \exp[t\, Q_G(u)]\, u^a\, du. \tag{3.17}$$

We can obtain the mgfs for several BG distributions from equations (3.16) and (3.17). For example, the mgfs of the BE (with parameter λ, for $t < \lambda^{-1}$) and BSL (for $t < 1$) distributions are

$$M(t) = \sum_{k=0}^{\infty}(a+k)\,B(k+a, 1-\lambda t)\,w_k,$$

and

$$M(t) = \sum_{k=0}^{\infty}(a+k)\,B(t+k+a, 1-t)\,w_k,$$

respectively. For the BPa distribution, the generating function can not be obtained from (3.16).

3.9. MEAN DEVIATIONS

For $X \sim$ beta-G$(a, b, \boldsymbol{\tau})$, the mean deviation $(\delta_1 = \mathrm{E}(|X - \mu_1'|))$ and the deviation about the median $(\delta_2(X) = \mathrm{E}(|X - M|))$ of X can be expressed as

$$\delta_1 = 2\mu_1'\,F(\mu_1') - 2m_1(\mu_1') \qquad \text{and} \qquad \delta_2 = \mu_1' - 2m_1(M), \qquad (3.18)$$

respectively, where $\mu_1' = \mathrm{E}(X)$, $M = \mathrm{Median}(X)$ is the median obtained from (3.3) by $M = Q(0.5)$, $F(\mu_1')$ is easily evaluated from the cdf (3.1) and $m_1(z) = \int_{-\infty}^{z} x\,f(x)\mathrm{d}x$ is the first incomplete moment of X.

In this section, we provide three alternative ways to compute δ_1 and δ_2. The first one is based on equation (3.2) and the binomial expansion

$$m_1(z) = \frac{1}{B(a,b)} \int_{-\infty}^{z} g(x)\,G(x)^{a-1}\,[1 - G(x)]^{b-1}dx$$

$$= \frac{1}{B(a,b)} \sum_{k=0}^{\infty}(-1)^k \binom{b-1}{k} \int_{0}^{z} x\,g(x)\,G(x)^{a+k-1}dx,$$

and then by integrating by parts

$$m_1(z) = \sum_{k=0}^{\infty} p_k(a,b)\,[z\,G(z)^{a+k} - H_k(z)], \qquad (3.19)$$

where

$$H_k(z) = \int_{0}^{z} G(x)^{a+k}dx, \quad p_k(a,b) = \frac{(-1)^k}{(a+k)\,B(a,b)} \binom{b-1}{k}.$$

So, the mean deviations in (3.18) can be computed from (3.19).

A second general equation for $m_1(z)$ can be derived from (3.6) as

$$m_1(z) = \sum_{k=0}^{\infty} w_k J_k(z), \tag{3.20}$$

where

$$J_k(z) = \int_{-\infty}^{z} x\, h_{a+k}(x) \mathrm{d}x. \tag{3.21}$$

Equation (3.21) is the basic quantity to determine the mean deviations of the exp-G models. Hence, the mean deviations (3.18) depend only on the mean deviations of the exp-G distribution. So, alternative expressions for δ_1 and δ_2 are

$$\delta_1 = 2\mu_1' F(\mu_1') - 2\sum_{k=0}^{\infty} w_k J_k(\mu_1') \qquad \text{and} \qquad \delta_2 = \mu_1' - 2\sum_{k=0}^{\infty} w_k J_k(M).$$

A simple application of (3.20) and (3.21) refers to the BW distribution. The EW distribution with power parameter $(a+k)$ has pdf (for $x > 0$) given by

$$h_{a+k}(x) = c\,(a+k)\,\beta^c\, x^{c-1}\, \exp\{-(\beta x)^c\}\,[1 - \exp\{-(\beta x)^c\}]^{a+k-1}$$

and then

$$J_k(z) = c\,(a+k)\,\beta^c \int_0^z x^c \exp\{-(\beta x)^c\}\,[1 - \exp\{-(\beta x)^c\}]^{a+k-1}\,\mathrm{d}x$$

$$= c\,(a+k)\,\beta^c \sum_{r=0}^{\infty} (-1)^r \binom{a+k-1}{r} \int_0^z x^c \exp\{-(r+1)(\beta x)^c\}\,\mathrm{d}x.$$

The previous integral can be given in terms of the incomplete gamma function and then the mean deviations for the BW distribution can be determined from

$$m_1(z) = \frac{1}{\beta} \sum_{k,r=0}^{\infty} \frac{(-1)^r\,(a+k)\,w_k}{(r+1)^{1+1/c}} \binom{a+k-1}{r} \gamma(1 + c^{-1}, (r+1)(\beta z)^c).$$

A second general formula for $m_1(z)$ can be derived by setting $u = G(x)$ in (3.6)

$$m_1(z) = \sum_{k=0}^{\infty} (a+k)\,w_k\, T_k(z), \tag{3.22}$$

where

$$T_k(z) = \int_0^{G(z)} Q_G(u)\, u^{a+k-1} du \qquad (3.23)$$

is a simple integral based on $Q_G(u)$.

In a similar way, the mean deviations for any BG distribution can be obtained from equations (3.22) and (3.23). For example, the mean deviations of the BE (with parameter λ), BSL and BPa (with parameter $\nu > 0$) can follow (by using the generalized binomial expansion) from the functions

$$T_k(z) = \lambda^{-1}\,\Gamma(a+k-1) \sum_{j=0}^{\infty} \frac{(-1)^j\,\{1 - \exp(-j\lambda z)\}}{(j+1)!\,\Gamma(a+k-1-j)},$$

$$T_k(z) = \frac{1}{\Gamma(k)} \sum_{j=0}^{\infty} \frac{(-1)^j\,\Gamma(a+k+j)\,\{1 - \exp(-jz)\}}{(j+1)!}$$

and

$$T_k(z) = \sum_{j=0}^{\infty} \sum_{r=0}^{j} \frac{(-1)^j}{(1-r\nu)} \binom{a+k}{j} \binom{j}{r} z^{1-r\nu},$$

respectively.

Applications of the first incomplete moment $m_1(q)$ can be used to find Bonferroni and Lorenz curves for a given probability π, which are given by $B(\pi) = m_1(q)/(\pi\,\mu_1')$ and $L(\pi) = m_1(q)/\mu_1'$, respectively, where $\mu_1' = E(X)$ and $q = Q(\pi)$ follows from equation (3.3).

3.10. ORDER STATISTICS

We obtain the density function of the ith order statistic $X_{i:n}$, say $f_{i:n}(x)$, in a random sample of size n from the BG distribution. We have

$$f_{i:n}(x) = \frac{f(x)}{B(i, n-i+1)} \sum_{j=0}^{n-i} (-1)^j \binom{n-i}{j} F(x)^{i+j-1},$$

for $i = 1, \ldots, n$. For any BG model defined from the parent functions $g(x)$ and $G(x)$, we can write

$$f_{i:n}(x) = \frac{g(x)\,G(x)^{a-1}\,\{1-G(x)\}^{b-1}}{B(a,b)\,B(i, n-i+1)} \sum_{j=0}^{n-i} (-1)^j \binom{n-i}{j} F(x)^{i+j-1}. \qquad (3.24)$$

Using the incomplete beta function expansion for b real non-integer, we obtain from (3.1)

$$F(x) = \frac{G(x)^a}{B(a,b)} \sum_{m=0}^{\infty} \frac{(1-b)_m\, G(x)^m}{(a+m)m!} = \sum_{m=0}^{\infty} g_m\, G(x)^{a+m},$$

where $g_m = \frac{(1-b)_m}{(a+m)m!B(a,b)}$.

We can write using equations (3.13) and (3.14)

$$F(x)^{i+j-1} = \left(\sum_{m=0}^{\infty} g_m\, G(x)^{a+m} \right)^{i+j-1} = \sum_{m=0}^{\infty} f_{i+j-1,m}\, G(x)^{(i+j-1)a+m},$$

where the coefficients $f_{i+j-1,m}$ are given by

$$f_{i+j-1,m} = (m\, g_0)^{-1} \sum_{r=1}^{m} [r(n+1) - m]\, g_r\, f_{i+j-1,m-r},$$

for $m \geq 1$, and $f_{i+j-1,0} = g_0^{i+j-1}$.

Further, we can rewrite (3.24) by expanding its binomial term as

$$f_{i:n}(x) = \sum_{m,r=0}^{\infty} \sum_{j=0}^{n-i} e_{m,r,j}\, h_{(i+j)a+m+r}, \tag{3.25}$$

where

$$e_{m,r,j} = \frac{(-1)^{j+r}\, f_{i+j-1,m}}{B(a,b)\, B(i, n-i+1)} \binom{b-1}{r} \binom{n-i}{j}.$$

Based on equation (3.25), the density function of the BG order statistics can be written as a linear combination of exp-G densities. So, some structural properties of $X_{i:n}$ can follow from those of the exp-G distribution.

3.11. RELIABILITY

We provide an expression for the reliability R when X_1 and X_2 are independent random variables with support in $\mathbb{R}^+$ and both of them follow the same BG distribution. Let $f_i(x; a_i, b_i, \boldsymbol{\tau})$ and $F_i(x; a_i, b_i, \boldsymbol{\tau})$ denote the density and cumulative functions of X_i with generator parameters (a_i, b_i), for $i = 1, 2$. By definition $R = P(X_2 < X_1) = \int_0^\infty f_1(x) F_2(x)\, \mathrm{d}x$, we have

$$R = \int_0^\infty f_1(x; a_1, b_1, \boldsymbol{\tau})\, F_2(x; a_2, b_2, \boldsymbol{\tau}) dx.$$

Then, equations (3.4) and (3.5) yield

$$f_1(x) = B(a_1, b_1)^{-1} f(x; a_1, b_1, \boldsymbol{\tau}) \sum_{i=0}^{\infty} (-1)^i \binom{b_1 - 1}{i} F_1(x; a_1, b_1, \boldsymbol{\tau})^{a_1 + i - 1}$$

and

$$F_2(x) = B(a_2, b_2)^{-1} \sum_{j=0}^{\infty} (-1)^j \binom{b_2 - 1}{j} (a_2 + j)^{-1} F_2(x; a_2, b_2, \boldsymbol{\tau})^{a_2 + j}.$$

Hence, after some algebra, we obtain

$$R = \sum_{i,j=0}^{\infty} \frac{\binom{b_1}{i} \binom{b_2}{j} [a_1(1 + i) + a_2(1 + j)]^{-1}}{a_2 (1 + i) B(a_1, b_1 + 1) B(a_2, b_2 + 1)}.$$

3.12. ENTROPY

The Shannon (1948) [19] entropy

$$H(f) = -\mathrm{E}_f[\log f(X; a, b, \boldsymbol{\tau})] = -\int f(x; a, b, \boldsymbol{\tau}) \log f(x; a, b, \boldsymbol{\tau}) \, \mathrm{d}x$$

(called simply entropy for short) is a measure of the uncertainty in a distribution, which implies that its negative is a measure of information. It was proved by Ebrahimi *et al.* (1999) [149] that there is no universal relationship between variance and entropy.

It can be proved that the BG density (3.2) with baseline G distribution satisfies

$$
\begin{aligned}
-\mathrm{E}_f[\log G(X; \boldsymbol{\tau})] &= \zeta(a, b), & (3.26) \\
-\mathrm{E}_f[1 - \log G(X; \boldsymbol{\tau})] &= \zeta(b, a), & (3.27) \\
\mathrm{E}_f[\log g(X; \boldsymbol{\tau})] - \mathrm{E}_U[\log g(G^{-1}(U; \boldsymbol{\tau}); \boldsymbol{\tau})] &= 0, & (3.28)
\end{aligned}
$$

where $\zeta(a, b) = \psi(a + b) - \psi(a)$, $\psi(\cdot)$ denotes the digamma function and $U \sim$ Beta(a, b). Further, the BG class has the maximum entropy of all distributions satisfying the information constraints (3.26)-(3.28). In fact, the entropy of X reduces to

$$
\begin{aligned}
-\mathrm{E}_f[\log f(X; a, b, \boldsymbol{\tau})] = &\log B(a, b) + (a - 1)\zeta(a, b) + (b - 1)\zeta(b, a) \\
&- \mathrm{E}_U[\log g(G^{-1}(U; \boldsymbol{\tau}); \boldsymbol{\tau})].
\end{aligned}
$$

So, the BG entropy is a sum of the entropy of the beta distribution and other term related to the entropy of the parent distribution:

$$\mathrm{E}_U[\log g(G^{-1}(U;\boldsymbol{\tau});\boldsymbol{\tau})].$$

Finally, equations (3.26) and (3.27) have information only about the beta model.

3.13. ESTIMATION

In this section, we obtain the MLEs of the parameters of the BG family. Let $x_1,\ldots,x_n$ be a sample of size n from $X \sim \mathrm{BG}(F;a,b,\boldsymbol{\tau})$, where $\boldsymbol{\tau} = (\tau_1,\ldots,\tau_p)^\top$ denotes the $p \times 1$ vector of unknown parameters of the baseline cdf $G(x) = G(x;\boldsymbol{\tau})$. The log-likelihood function for $\boldsymbol{\theta}$ is

$$\begin{aligned}
\ell(\boldsymbol{\theta}) = & \, n\log B(a,b) + \sum_{i=1}^{n}\log g(x_i) + (a-1)\sum_{i=1}^{n}\log G(x_i) \\
& + (b-1)\sum_{i=1}^{n}\log[1-G(x_i)].
\end{aligned} \tag{3.29}$$

The function (3.29) can be maximized either directly by using SAS (Proc NLMixed) or the Ox (sub-routine MaxBFGS) (Doornik, 2007 [150]) or by solving the nonlinear likelihood equations by differentiating $\ell(\boldsymbol{\theta})$.

Initial estimates of the parameters a and b may be inferred from the estimate of $\boldsymbol{\tau}$. To see why, note that $X \sim \mathrm{BG}(a,b,\boldsymbol{\tau})$ gives $G(X;\boldsymbol{\tau}) \sim \mathrm{Beta}(a,b)$. Then, knowing the moments of the BG family, we have

$$\mathrm{E}[G(X)^s] = \frac{B(a+s,b)}{B(a,b)}, \tag{3.30}$$

$$\mathrm{E}\{[1-G(X)]^r\} = \frac{B(a,b+r)}{B(a,b)},$$

$$\mathrm{E}\{G(X)^s[1-G(X)]^r\} = \frac{B(a+s,b+r)}{B(a,b)}. \tag{3.31}$$

Equation (3.30) with $s=1$ and (3.31) with $s=1$ and $r=1$ give

$$\mathrm{E}[G(X)] = \frac{a}{a+b},$$

$$\mathrm{E}\{G(X)[1-G(X)]\} = \frac{ab}{(a+b)(a+b+1)}.$$

Solving the last two equations for a and b yields

$$a = \frac{u\,v}{u\,(1-u) - v} \quad \text{and} \quad b = \frac{v\,(1-u)}{u\,(1-u) - v},$$

where $u = E[G(X)]$ and $v = E\{G(X)\,[1 - G(X)]\}$.

The components of the score vector $\boldsymbol{U}(\boldsymbol{\theta}) = (U_a(\boldsymbol{\theta}), U_b(\boldsymbol{\theta}), \boldsymbol{U}_{\boldsymbol{\tau}}(\boldsymbol{\theta})^{\top})^{\top}$ are obtained by differentiating the log-likelihood function with respect to the parameters. We have

$$U_a(\boldsymbol{\theta}) = -n\big[\psi(a) - \psi(a+b)\big] + \sum_{i=1}^{n} \log[G(x_i)],$$

$$U_b(\boldsymbol{\theta}) = -n\big[\psi(b) - \psi(a+b)\big] + \sum_{i=1}^{n} \log[1 - G(x_i)],$$

$$\boldsymbol{U}_{\boldsymbol{\tau}}(\boldsymbol{\theta}) = \sum_{i=1}^{n} \frac{\dot{g}(x_i)_{\boldsymbol{\tau}}}{g(x_i)} + (a-1) \sum_{i=1}^{n} \frac{\dot{G}(x_i)_{\boldsymbol{\tau}}}{G(x_i)} + (b-1) \sum_{i=1}^{n} \frac{\dot{G}(x_i)_{\boldsymbol{\tau}}}{1 - G(x_i)},$$

where $\dot{g}(x_i)_{\boldsymbol{\tau}} = \partial g(x_i)/\partial \boldsymbol{\tau}$ and $\dot{G}(x_i)_{\boldsymbol{\tau}} = \partial G(x_i)/\partial \boldsymbol{\tau}$ are $p \times 1$ vectors. The MLEs of the unknown parameters are obtained from

$$U_a(\boldsymbol{\theta}) = 0, \quad U_b(\boldsymbol{\theta}) = 0, \quad \boldsymbol{U}_{\boldsymbol{\tau}}(\boldsymbol{\theta}) = \boldsymbol{0},$$

and then by solving these equations numerically using iterative Newton–Raphson type algorithms.

3.14.   BETA-G REGRESSION MODEL

Ortega *et al.* (2011) [151] pioneered the log-beta-Weibull (log-BW) regression model based on the BW distribution. This regression model includes as special cases well-known models that can be applied to censored survival data.

The Neyman type A beta Weibull model with long-term survivors was introduced by Hashimoto *et al.* (2013) [152] with an application in medical field. Recently, two papers considered regression models using the BW distribution. Ortega *et al.* (2012) [153] defined the negative binomial-beta Weibull regression model for estimating the recurrence probability of prostate cancer and the cure fraction for patients with clinically localized prostate cancer treated by radical prostatectomy.

Ortega *et al.* (2012) [154] defined the log-beta-Birnbaum-Saunders distribution by taking the logarithm of the BBS distribution. They obtained explicit expressions for its moments and mgf. Further, the log-beta Birnbaum-Saunders (LBBS) regression model was introduced by these authors. This regression model can be applied to censored data and be used more effectively in survival analysis. More recently, Cordeiro *et al.* (2013) [131] defined the log-beta exponentiated Weibull regression model to analyze censored data and Ortega *et al.* (2013) [155] studied some properties of the log-beta Weibull (LBW) distribution and derived formal expressions for the moments, mgf, qf and mean deviations.

3.14.1. An Extended Weibull Distribution

Most extended Weibull distributions have been explored in reliability to provide better fits to real data sets than the classic two or three parameter Weibull models. A complex extension of the Weibull model with more than three parameters usually reduces the probability of interpreting the parameters. Xie *et al.* (2002) [156] proposed a generalization of the Weibull model with three parameters λ, α and τ, so-called the extended Weibull distribution, with pdf defined by (for $t \geq 0$)

$$f(t; \lambda, \tau, \alpha) = \lambda \tau \left(\frac{t}{\alpha}\right)^{\tau-1} \exp\left\{\left(\frac{t}{\alpha}\right)^{\tau} + \lambda \alpha \left[1 - \exp\left(\left(\frac{t}{\alpha}\right)^{\tau}\right)\right]\right\}, \quad (3.32)$$

where $\lambda > 0$ and $\alpha > 0$ are scale parameters and $\tau > 0$ is a shape parameter. The corresponding survival and failure rate functions are

$$S(t; \lambda, \tau, \alpha) = \exp\left\{\lambda \alpha \left[1 - \exp\left\{\left(\frac{t}{\alpha}\right)^{\tau}\right\}\right]\right\}$$

and

$$h(t; \lambda, \tau, \alpha) = \lambda \tau \left(\frac{t}{\alpha}\right)^{\tau-1} \exp\left[\left(\frac{t}{\alpha}\right)^{\tau}\right],$$

respectively.

The extended Weibull distribution can be simulated by the inversion method. If U is a uniform random variable in the interval $(0,1)$, then the random variable $T = \{\alpha^{\tau} \log[1 - (\lambda\alpha)^{-1} \log(1 - U)]\}^{\frac{1}{\tau}}$ follows an extended Weibull random variable with parameters λ, τ and α.

3.14.2. The Log-extended Weibull Distribution

Let T be a random variable having the extended Weibull density (3.32). We define the random variable $Y = \log(T)$ having a log-extended Weibull distribution parametrized in terms of σ $(-\tau^{-1})$ and μ $(= \log(\alpha))$ by the pdf (for $y \in \mathbb{R}$)

$$f(y; \lambda, \sigma, \mu) = \frac{\lambda}{\sigma} \exp\left(\frac{y-\mu}{\sigma}\right)$$

$$\times \exp\left\{\mu + \exp\left(\frac{y-\mu}{\sigma}\right) + \lambda \exp(\mu)\left[1 - \exp\left[\exp\left(\frac{y-\mu}{\sigma}\right)\right]\right]\right\},$$

where $\lambda > 0$, $\sigma > 0$ and $\mu \in \mathbb{R}$. The corresponding survival function reduces to

$$S(y; \lambda, \sigma, \mu) = \exp\left\{\lambda \exp(\mu)\left[1 - \exp\left[\exp\left(\frac{y-\mu}{\sigma}\right)\right]\right]\right\}.$$

We define the random variable $Z = (Y - \mu)/\sigma$ with density function (for $z \in \mathbb{R}$)

$$f(z; \lambda, \mu) = \lambda \exp\{z + \mu + \exp(z) + \lambda \exp(\mu)[1 - \exp[\exp(z)]]\}. \quad (3.33)$$

3.14.3. The Log-extended Weibull Regression Model

In many applications, the lifetimes are functions of explanatory variables such as soil quality, climate, blood pressure, socioeconomic and school variables, among many others. Let $\boldsymbol{x}_i^\top = (x_{i1}, \ldots, x_{ip})$ be the explanatory variable vector related to the ith response variable y_i. We define a linear regression model linking the response variable Y_i and the vector $\boldsymbol{x}_i$ by

$$Y_i = \boldsymbol{x}_i^\top \boldsymbol{\beta} + \sigma Z_i, \ i = 1, \ldots, n, \quad (3.34)$$

where the random error Z_i has the distribution (3.33), $\boldsymbol{\beta} = (\beta_1, \ldots, \beta_p)^\top$ and $\sigma > 0$ and $\lambda > 0$ are unknown parameters. The location parameter vector $\boldsymbol{\mu} = (\mu_1, \ldots, \mu_n)^\top$ of the log-extended Weibull (LEW) regression model is given by a linear model $\mu = \mathbf{X}\boldsymbol{\beta}$, where $\mathbf{X} = (\boldsymbol{x}_1, \ldots, \boldsymbol{x}_n)^\top$ is a known model matrix. Henceforth, the log-Weibull (LW) (or extreme value) regression model is given by (3.34) when $\alpha \to \infty$.

Let $(y_1, \boldsymbol{x}_1), \ldots, (y_n, \boldsymbol{x}_n)$ be n independent observations, where each random response is defined by $y_i = \min\{\log(t_i), \log(c_i)\}$. We consider non-informative censoring and that the observed lifetimes and censoring times are independent. Let F and C be the sets of individuals for which y_i is the log-lifetime or log-censoring, respectively. The total log-likelihood function for $\boldsymbol{\theta}$ follows from (3.33) and (3.34) as

$$
\begin{aligned}
\ell(\boldsymbol{\theta}) \;=\; & r\log(\lambda) - r\log(\sigma) + \sum_{i \in F} \boldsymbol{x}_i^\top \boldsymbol{\beta} + \sum_{i \in F} z_i + \sum_{i \in F} \exp(z_i) \\
& + \sum_{i \in F} \lambda \exp(\boldsymbol{x}_i^\top \boldsymbol{\beta}) \Big\{ 1 - \exp[\exp(z_i)] \Big\} \\
& + \sum_{i \in C} \lambda \exp(\boldsymbol{x}_i^\top \boldsymbol{\beta}) \Big\{ 1 - \exp[\exp(z_i)] \Big\},
\end{aligned}
\tag{3.35}
$$

where r is the number of uncensored observations (failures) and $z_i = (y_i - \boldsymbol{x}_i^\top \boldsymbol{\beta})/\sigma$. The MLE $\widehat{\boldsymbol{\theta}}$ of $\boldsymbol{\theta}$ can be determined by maximizing the log-likelihood function (3.35). We evaluate $\widehat{\boldsymbol{\theta}}$ using Ox (MaxBFGS subroutine) (see Doornik, 2007) [150]. After fitting model (3.34), the survival function of Y_i can be estimated by

$$
S(y_i; \widehat{\boldsymbol{\beta}}^\top, \widehat{\lambda}, \widehat{\sigma}) \;=\; \exp\left[\widehat{\lambda} \exp(\boldsymbol{x}_i^\top \widehat{\boldsymbol{\beta}}) \left\{ 1 - \exp\left[\exp\left(\frac{y - \boldsymbol{x}_i^\top \widehat{\boldsymbol{\beta}}}{\widehat{\sigma}} \right) \right] \right\} \right].
$$

Note that the asymptotic distribution of $(\widehat{\boldsymbol{\theta}} - \boldsymbol{\theta})$ is multivariate normal under standard regularity conditions, $N_{p+2}(0, K(\boldsymbol{\theta})^{-1})$, where $K(\boldsymbol{\theta})$ is the expected information matrix. The asymptotic covariance matrix $K(\boldsymbol{\theta})^{-1}$ of $\widehat{\boldsymbol{\theta}}$ can be approximated by the inverse of the observed information matrix $-\ddot{L}(\boldsymbol{\theta})$, whose elements can be evaluated numerically. So, the asymptotic inference for the parameter vector $\boldsymbol{\theta}$ can be based on the multivariate normal $N_{p+2}(0, -\ddot{L}(\boldsymbol{\theta})^{-1})$. Hence, an $100(1 - \gamma)\%$ asymptotic confidence interval for the parameter θ_r is given by

$$
ACI_r = \left(\hat{\theta}_r - z_{\gamma/2} \sqrt{-\widehat{\ddot{L}}^{r,r}}, \; \hat{\theta}_r + z_{\gamma/2} \sqrt{-\widehat{\ddot{L}}^{r,r}} \right),
$$

where $-\widehat{\ddot{L}}^{r,r}$ denotes the rth diagonal element of $-\ddot{L}(\widehat{\boldsymbol{\theta}})^{-1}$ and $z_{\gamma/2}$ is the quantile $1 - \gamma/2$ of the standard normal distribution.

3.15.   CONCLUSIONS

In the chapter, we study a family of beta-G (BG) distributions, which can include as special models all continuous distributions. For any baseline continuous distribution G, we define the associated BG distribution with two extra positive parameters. The new family extends several well-known distributions like the normal, Weibull, Gumbel, gamma and log-normal distributions. We investigate some of its mathematical properties such as the moments, generating and quantile functions, mean deviations, Bonferroni and Lorenz curves, Shannon entropy, and order statistics. We estimate the model parameters by the maximum likelihood method and define a regression model based on an extended Weibull distribution.

Chapter 4

Special Generalized Beta Models

Abstract: In this chapter, we provide theoretical essays about five special models in the beta family. For each model, its cdf, pdf and hrf have explicit forms, which can be utilized for determining some mathematical properties. Two applications are performed in order to illustrate the flexibility of the densities under discussion.

Keywords:: BBS; Beta-G model; BF; BGE; BMW; BW; Moment.

This chapter includes a discussion about five special beta models: beta generalized exponential, beta Weibull, beta Fréchet, beta modified Weibull and beta Birnbaum-Saunders distributions. Two applications to real data with positive support emphasize the flexibility of these five models.

4.1. BETA GENERALIZED EXPONENTIAL

As mentioned in Chapter 2, a random variable T is said to have the generalized exponential (GE) distribution if its cdf and pdf are

$$G(x; \lambda, \alpha) = (1 - e^{-\lambda x})^\alpha \quad \text{and} \quad g(x; \lambda, \alpha) = \alpha\,\lambda\,e^{-\lambda x}(1 - e^{-\lambda x})^{\alpha-1}, \quad (4.1)$$

respectively. The shape ($\alpha > 0$) and scale ($\lambda > 0$) parameters of the GE distribution are similar to those of the gamma and Weibull distributions.

The distribution (4.1) is also named the exponentiated exponential (EE) distribution. Note that the exponential distribution is a special case of the GE

Gauss M. Cordeiro, Rodrigo B. Silva & Abraão D. C. Nascimento

distribution when $\alpha = 1$.

The four-parameter beta generalized exponential (BGE) distribution is defined by taking $G(x)$ in equation (4.1) as the baseline cdf in (3.1). Thus, a random variable X is said to have the BGE distribution if its cdf and pdf are (for $x > 0$)

$$F(x; a, b, \lambda, \alpha) = I_{(1-e^{-\lambda x})^\alpha}(a, b) \qquad (4.2)$$

and

$$f(x; a, b, \lambda, \alpha) = \frac{\alpha\lambda}{B(a,b)} e^{-\lambda x} (1 - e^{-\lambda x})^{\alpha a - 1} \left\{ 1 - (1 - e^{-\lambda x})^\alpha \right\}^{b-1}, \qquad (4.3)$$

respectively, for $a, b, \lambda, \alpha > 0$. The corresponding hrf becomes

$$\tau(x; \lambda, \alpha) = \frac{\alpha\lambda e^{-\lambda x} (1 - e^{-\lambda x})^{\alpha a - 1} \left\{ 1 - (1 - e^{-\lambda x})^\alpha \right\}^{b-1}}{B(a,b)[1 - I_{(1-e^{-\lambda x})^\alpha}(a, b)]}. \qquad (4.4)$$

Note that the pdf (4.3) does not involve any complicated function. If X is a random variable with pdf (4.3), we write $X \sim \text{BGE}(a, b, \lambda, \alpha)$. The BGE distribution generalizes some well-known distributions in the literature. The GE distribution is a special case when $a = b = 1$. For $\alpha = 1$, we obtain the exponential distribution with parameter λ. The BE distribution follows from (4.2) when $\alpha = 1$. The hrf (4.4) can be bathtub shaped, monotonically increasing or decreasing and upside-down bathtub depending on the parameter selection.

There are two simple formulae for the cdf of the BGE distribution depending if the parameter $b > 0$ is real non-integer or integer. Note that, if $|z| < 1$ and $b > 0$ is real non-integer, we have

$$(1 - z)^{b-1} = \sum_{j=0}^{\infty} \frac{(-1)^j \Gamma(b)}{\Gamma(b - j)j!} z^j, \qquad (4.5)$$

where $\Gamma(\cdot)$ is the gamma function. Using the expansion (4.5) in (4.2), the cdf of the BGE distribution when $b > 0$ is real non-integer follows as

$$F(x; a, b, \alpha, \lambda) = \frac{\Gamma(b)}{B(a,b)} \sum_{j=0}^{\infty} \frac{(-1)^j}{\Gamma(b - j)j!} \int_0^{(1-e^{-\lambda x})^\alpha} \omega^{a+j-1} d\omega$$

$$= \frac{\Gamma(a+b)}{\Gamma(a)} \sum_{j=0}^{\infty} \frac{(-1)^j G(x; \lambda, \alpha(a + j))^{\alpha(a+j)}}{\Gamma(b - j)j!(a + j)}$$

$$= \sum_{j=0}^{\infty} w_j \, G(x; \lambda, \alpha(a + j))^{\alpha(a+j)}, \qquad (4.6)$$

where $w_j = (-1)^j \Gamma(a+b)[(a+j)\,\Gamma(a)\,\Gamma(b-j)\,j!]^{-1}$. Equation (4.6) reveals that the cdf of the BGE distribution can be expressed as an infinite weighted sum of cdfs of GE distributions. The BE cdf follows with $\alpha = 1$ from (4.6). The cdf of the double generalized exponential (DGE) distribution follows from (4.6) when $a = 1$.

By differentiating (4.6), the density function (4.3) can be expressed as a linear combination of the GE pdfs

$$f(x; a, b, \lambda, \alpha) = \sum_{j=0}^{\infty} w_j\, f^*(x; \lambda, \alpha\,(a+j)),$$

where $f^*(x; \lambda, \alpha\,(a+j)) = \alpha(a+j)\,g(x; \lambda, \alpha\,(a+j))\,G(x; \lambda, \alpha\,(a+j))^{\alpha(a+j)-1}$ is the GE density function corresponding to the cdf $G(x; \lambda, \alpha\,(a+j))^{\alpha(a+j)}$ in equation (4.6). Thus, the BGE distribution has the advantage that some of its mathematical properties can be directly obtained from the corresponding properties of the GE distribution, such as those explored in Sections 3.5 and 3.8.

4.2. BETA WEIBULL

A random variable T is said to have the Weibull distribution if its cdf and pdf are

$$G(x; \alpha, \beta) = 1 - \mathrm{e}^{-(\beta x)^\alpha} \quad \text{and} \quad g(x; \alpha, \beta) = \alpha\,\beta^\alpha x^{\alpha-1}\mathrm{e}^{-(\beta x)^\alpha}, \quad x > 0,$$

respectively. The four-parameter beta Weibull (BW) cdf is defined by inserting the above $G(x; \alpha, \beta)$ in equation (3.1). Thus, a random variable X is said to have the BW distribution if its cdf and pdf are (for $x > 0$)

$$F(x; a, b, \alpha, \beta) = I_{1-\exp\{-(\beta x)^\alpha\}}(a, b) \tag{4.7}$$

and

$$f(x; a, b, \alpha, \beta) = \frac{\alpha\,\beta^\alpha}{B(a, b)} x^{\alpha-1} \exp\{-b\,(\beta x)^\alpha\}\,[1 - \exp\{-(\beta x)^\alpha\}]^{a-1}, \tag{4.8}$$

respectively, for $a > 0$, $b > 0$, $\beta > 0$ and $\alpha > 0$. The associated hrf is

$$\tau(x; a, b, \alpha, \beta) = \frac{\alpha\,\beta^\alpha \exp\{-b\,(\beta x)^\alpha\}\,[1 - \exp\{-(\beta x)^\alpha\}]^{a-1}}{B(a, b)I_{1-\exp\{-(\beta x)^\alpha\}}(a, b)}.$$

If X is a random variable with pdf (4.8), we write $X \sim BW(a, b, \alpha, \beta)$. The BW distribution contains as special case the EW distribution when $b = 1$.

The Weibull distribution (with parameters α and β) is clearly a basic exemplar when $a = b = 1$. For $a = 1$, the density (4.8) follows a Weibull distribution with parameters $\beta \, b^{1/\alpha}$ and α. The beta exponential distribution is also a special case for $\alpha = 1$.

Cordeiro *et al.* (2011) [90] obtained closed-form expressions for the distribution and density functions of the BW distribution. Starting from the explicit expression for the cdf (4.7)

$$F(x; a, b, \alpha, \beta) = \frac{\alpha \, \beta^\alpha}{B(a, b)} \int_0^x y^{\alpha-1} \exp\left\{-b\,(\beta y)^\alpha\right\} [1 - \exp\left\{-(\beta y)^\alpha\right\}]^{a-1} dy,$$

and changing variables $(\beta y)^\alpha = u$ gives

$$F(x; a, b, \alpha, \beta) = \frac{1}{B(a, b)} \int_0^{(\beta x)^\alpha} e^{-bu}(1 - e^{-u})^{a-1} du.$$

Using expansion (4.5) in the above equation, we have

$$F(x; a, b, \alpha, \beta) = \frac{\Gamma(a + b)}{\Gamma(b)} \sum_{j=0}^{\infty} \frac{(-1)^j \left\{1 - e^{-(b+j)(\beta x)^\alpha}\right\}}{\Gamma(a - j)j!(b + j)}, \quad x > 0, \qquad (4.9)$$

if $a > 0$ is real non-integer. In this case, the above expansion may be used for further analytical or numerical studies.

4.3. BETA FRÉCHET

A random variable T is said to have the Fréchet distribution if its cdf and pdf (for $x > 0$) are

$$G(x; \sigma, \lambda) = \exp\left\{-\left(\frac{\sigma}{x}\right)^\lambda\right\}$$

and

$$g(x; \sigma, \lambda) = \lambda \, \sigma^\lambda x^{-(\lambda+1)} \exp\left\{-\left(\frac{\sigma}{x}\right)^\lambda\right\},$$

respectively, where $\sigma > 0$ is the scale parameter and $\lambda > 0$ is the shape parameter. The cdf of the beta Fréchet (BF) distribution with parameters $a > 0$, $b > 0$, $\sigma > 0$ and $\lambda > 0$ is determined by applying the above cdf $G(x; \sigma, \lambda)$ to equation (3.1), which gives (for $x > 0$)

$$F(x; a, b, \sigma, \lambda) = \int_0^{\exp\left\{-\left(\frac{\sigma}{x}\right)^\lambda\right\}} \omega^{a-1}(1 - \omega)^{b-1} d\omega. \qquad (4.10)$$

The corresponding density and hazard functions are (for $x > 0$)

$$f(x; a, b, \sigma, \lambda) = \frac{\lambda \sigma^\lambda \exp\left\{-a\left(\frac{\sigma}{x}\right)^\lambda\right\} \left[1 - \exp\left\{-\left(\frac{\sigma}{x}\right)^\lambda\right\}\right]^{b-1}}{x^{\lambda+1} B(a, b)} \qquad (4.11)$$

and

$$\tau(x; a, b, \sigma, \lambda) = \frac{\lambda \sigma^\lambda \exp\left\{-a\left(\frac{\sigma}{x}\right)^\lambda\right\} \left[1 - \exp\left\{-\left(\frac{\sigma}{x}\right)^\lambda\right\}\right]^{b-1}}{x^{\lambda+1} B(a, b) \, I_{1-\exp\left\{-\left(\frac{\sigma}{x}\right)^\lambda\right\}}(a, b)},$$

respectively.

If X is a random variable with pdf (4.11), we write $X \sim \mathrm{BF}(a, b, \sigma, \lambda)$. The Fréchet distribution is a special case of the BF distribution when $a = b = 1$. The exponentiated Fréchet (EF) distribution is also a special case when $a = 1$. Further, for $b = 1$ and $\lambda = 1$, (4.11) is an inverse gamma distribution with shape parameter two and scale parameter $a\sigma$.

Based on equation (4.5), Barreto-Souza *et al.* (2011) [111] provided some expansions for the distribution and density functions depending on whether the parameter b (or a) is real non-integer or integer. Application of (4.5) to (4.10), if b is real non-integer, gives

$$F(x; a, b, \sigma, \lambda) = \frac{\Gamma(a + b)}{\Gamma(a)} \sum_{j=0}^{\infty} \frac{(-1)^j \exp\{-(a + j)\left(\frac{\sigma}{x}\right)^\lambda\}}{\Gamma(b - j) j! (a + j)}. \qquad (4.12)$$

For b integer, the sum in (4.12) simply stops at $b - 1$.

If $b > 0$ is real non-integer, and again using (4.5), the BF density function can be rewritten as

$$f(x; a, b, \sigma, \lambda) = \sum_{k=0}^{\infty} v_k \, g(x; \alpha_k, \lambda), \qquad (4.13)$$

where

$$v_k = \frac{(-1)^k \, \Gamma(a + b)}{(a + k) \, k! \, \Gamma(a) \Gamma(b - k)}$$

(for $k = 0, 1, \dots$) are coefficients such that $\sum_{k=0}^{\infty} v_k = 1$ and $g(x; \alpha_k, \lambda)$ is the Fréchet density with scale parameter $\alpha_k = \sigma(k + a)^{1/\lambda}$ and shape parameter λ. Thus, some properties of the BF distribution can be directly obtained from (4.13) and the corresponding properties of the Fréchet distribution.

4.4. BETA MODIFIED WEIBULL

The modified Weibull (MW) distribution having three parameters $\gamma > 0$, $\alpha > 0$ and $\lambda \geq 0$ has cdf and pdf given by (for $x > 0$)

$$G(x; \gamma, \alpha, \lambda) = 1 - \exp\{-\alpha x^\gamma \exp(\lambda x)\} \qquad (4.14)$$

and
$$g(x; \gamma, \alpha, \lambda) = \alpha \, x^{\gamma-1}(\gamma + \lambda x) \exp\{\lambda x - \alpha x^{\gamma} \exp(\lambda x)\},$$

respectively. Note that α and γ are the scale and shape parameters of the MW distribution, respectively. The parameter λ represents a fragility factor in the survival function of the individuals as the time increases. The Weibull distribution is a special case of (4.14) when $\lambda = 0$. In this case, if $\gamma = 1$ and $\gamma = 2$, we obtain the exponential and Rayleigh distributions, respectively.

Silva *et al.* (2010) [85] introduced the beta modified Weibull (BMW) distribution by inserting $G(x)$ in (4.14) in equation (3.1). Hence, the general form for the BMW cdf is (for $x > 0$)

$$F(x; a, b, \gamma, \alpha, \lambda) = \int_0^{1-\exp\{-\alpha x^{\gamma} \exp(\lambda x)\}} \omega^{a-1}(1-\omega)^{b-1} d\omega.$$

The corresponding density and hazard rate functions are

$$f(x; a, b, \gamma, \alpha, \beta) = \frac{\alpha \, x^{\gamma-1}(\gamma + \lambda x) \exp(\lambda x)}{B(a,b)} [1 - \exp\{-\alpha \, x^{\gamma} \exp(\lambda x)\}]^{a-1}$$
$$\times \exp\{-b\alpha \, x^{\gamma} \exp(\lambda x)\} \tag{4.15}$$

and

$$\tau(x; a, b, \gamma, \alpha, \beta) = \frac{\alpha \, x^{\gamma-1}(\gamma + \lambda x) \exp(\lambda x)}{B(a,b)[1 - I_{1-\exp\{-\alpha x^{\gamma} \exp(\lambda x)\}}(a,b)]}$$
$$\times [1 - \exp\{-\alpha \, x^{\gamma} \exp(\lambda x)\}]^{a-1}$$
$$\times \exp\{-b\alpha \, x^{\gamma} \exp(\lambda x)\},$$

respectively. If X is a random variable with density (4.15), we write $X \sim$ BMW$(a, b, \gamma, \alpha, \beta)$.

The BMW distribution includes, as special cases, some well-known distributions. For example, it reduces to the BW distribution when $\lambda = 0$. In this case, if $\gamma = 1$, it becomes the BE distribution. The generalized modified Weibull (GMW) distribution is also a special case when $b = 1$. For $a = b = 1$, the special case refers to the MW distribution. For $b = 1$ and $\lambda = 0$, the BMW distribution is identical to the EW distribution. If $\gamma = b = 1$ and $\lambda = 0$, the BMW distribution coincides with the EE distribution. For $\gamma = 2$, $\lambda = 0$ and $b = 1$, we have the generalized Rayleigh (GR) distribution. The Weibull distribution is clearly a special case for $a = b = 1$ and $\lambda = 0$. A desirable property of the BMW distribution is that its hrf can accommodate four different forms: bathtub shaped, monotonically increasing or decreasing and upside-down bathtub.

Silva *et al.* (2010) [85] demonstrated that the BMW density can be expressed as a linear combination of MW densities. This result is essential to obtain some mathematical properties of the BMW distribution from the properties of the MW distribution. For $a > 0$ real non-integer, they demonstrated that (for $x > 0$)

$$F(x; a, b, \gamma, \alpha, \beta) = \sum_{j=0}^{\infty} q_j \, G(x; \gamma, \alpha(b+j), \lambda),$$

where $G(x; \gamma, \alpha(b+j), \lambda)$ is the cdf of the MW distribution with scale parameter $\alpha(b+j)$, shape parameter γ and accelerated parameter λ, and

$$q_j = q_j(a, b) = \frac{(-1)^j \, \Gamma(a)}{(b+j) \, j! \, B(a, b) \Gamma(a-j)},$$

for $j = 0, 1, 2, \ldots$, are constants such that $\sum_{j=0}^{\infty} q_j = 1$. If a is an integer, the index j in the previous sum stops at $a - 1$. The BMW density follows immediately as

$$f(x; a, b, \gamma, \alpha, \beta) = \sum_{j=0}^{\infty} q_j \, g(x; \gamma, \alpha(b+j), \lambda), \quad x > 0,$$

where $g(x; \gamma, \alpha(b+j), \lambda)$ is the pdf of the MW distribution with scale parameter $\alpha(b+j)$, shape parameter γ and accelerated parameter λ. Some properties of the BMW distribution can follow from the last equation and those of the MW distribution.

4.5. BETA BIRNBAUM-SAUNDERS

A random variable T has a Birnbaum-Saunders (BS) distribution, if it can be expressed as

$$T = \beta \left[\frac{\alpha Z}{2} + \left\{ \left(\frac{\alpha Z}{2} \right)^2 + 1 \right\}^{1/2} \right]^2,$$

where Z is a standard normal random variable. Its cdf is defined by

$$G(x; \alpha, \beta) = \Phi(v), \quad x > 0, \tag{4.16}$$

where $v = \alpha^{-1} \rho(x/\beta)$, $\rho(z) = z^{1/2} - z^{-1/2}$ and $\Phi(\cdot)$ is the standard normal distribution function. The pdf corresponding to (4.16) is

$$g(x; \alpha, \beta) = \kappa(\alpha, \beta) x^{-3/2} (x + \beta) \exp \left\{ -\frac{\tau(x/\beta)}{2\alpha^2} \right\}, \quad x > 0,$$

where $\kappa(\alpha, \beta) = \exp(\alpha^{-2})/(2\alpha\sqrt{2\pi\beta})$ and $\tau(z) = z + z^{-1}$.

We omit the dependence of the cdf and pdf on the parameters. The four-parameter beta Birnbaum-Saunders (BBS) cdf is obtained by applying (4.16) to equation (3.1), which gives

$$F(x) = I_{\Phi(v)}(a, b) = \int_0^{\Phi(v)} \omega^{a-1}(1 - \omega)^{b-1}d\omega, \quad x > 0. \qquad (4.17)$$

The corresponding density and hazard rate functions are (for $x > 0$)

$$f(x) = \frac{\kappa(\alpha, \beta)}{B(a, b)} x^{-3/2}(x + \beta) \exp\left\{-\tau(x/\beta)/(2\alpha^2)\right\} \Phi(v)^{a-1}[1 - \Phi(v)]^{b-1}, \qquad (4.18)$$

and

$$\tau(x) = \frac{\kappa(\alpha, \beta)x^{-3/2}(x + \beta) \exp\left\{-\tau(x/\beta)/(2\alpha^2)\right\} \Phi(v)^{a-1}[1 - \Phi(v)]^{b-1}}{B(a, b)[1 - I_{\Phi(v)}(a, b)]}.$$

If X is a random variable with density function defined in (4.18), we denote $X \sim \text{BBS}(a, b, \alpha, \beta)$. Cordeiro and Lemonte (2011) [118] provided expansions for the cdf of the BBS distribution in terms of infinite (or finite if both a and b are integers) weighted sums of powers of $\Phi(v)$ given by (4.16).

If $b > 0$ is real non-integer, by expanding $(1 - \omega)^{b-1}$ in (4.17) in power series gives

$$F(x) = \frac{1}{B(a, b)} \sum_{j=0}^{\infty} \omega_j \, \Phi(v)^{a+j}, \qquad (4.19)$$

where $\omega_j = (-1)^j \binom{b-1}{j}(a + j)^{-1}$ and the binomial coefficient is defined for any real $b > 0$. If b is an integer, the index j in (4.19) stops at $b - 1$.

If a is an integer, equation (4.19) gives the cdf of the BBS distribution as a power series of $\Phi(v)$. Otherwise, if a is positive real non-integer, we can expand $\Phi(v)^a$ as

$$\Phi(v)^a = \sum_{j=0}^{\infty} s_j(a) \, \Phi(v)^j, \qquad (4.20)$$

where

$$s_j(a) = \sum_{r=j}^{\infty} (-1)^{j+r} \binom{a}{r}\binom{r}{j}$$

and then $F(x)$ can be expressed from (4.19) and (4.20) as

$$F(x) = \frac{1}{B(a, b)} \sum_{j=0}^{\infty} t_j \Phi(v)^j, \qquad (4.21)$$

where $t_j = \sum_{m=0}^{\infty} \omega_m s_j(a+m)$. Expansions for the BBS density function follow immediately by simple differentiation of equations (4.19) and (4.21)

$$f(x) = \frac{g(x)}{B(a,b)} \sum_{j=0}^{\infty} (a+j)\omega_j \, \Phi(v)^{a+j-1} \qquad (4.22)$$

and

$$f(x) = \frac{g(x)}{B(a,b)} \sum_{j=0}^{\infty} (j+1)t_{j+1} \, \Phi(v)^j, \qquad (4.23)$$

respectively. Expansions (4.22) and (4.23) hold for $a > 0$ integer and a positive real non-integer, respectively. Equations (4.22)–(4.23) reveal that some properties of the BBS distribution can be determined from those exp-BS properties.

4.6. APPLICATIONS

We compare the fits of the BMW, BW, MW and EW distributions by means of a real data set. The model parameters are estimated by maximum likelihood using the NLMixed in SAS subroutine. We consider the lifetimes of 50 components given by Aarset [157], which possess a bathtub-shaped failure rate shape.

In Table 4.1, we list the MLEs of the parameters with corresponding standard errors (SEs) in parentheses, and the AIC and BIC statistics for the BMW, BW, MW and EW models.

Except for the EW distribution, all the competing models are adequate to model these data. However, the AIC and BIC statistics indicate that the BMW distribution outperforms the other distributions. Plots of the estimated pdfs and cdfs of the fitted BMW, BW, MW and EW models to the Aarset's data are displayed in Figure 4.1 by confirming the superiority of the BMW model.

Another example considers a real data set on the strengths of 1.5 cm glass fibres, measured at the National Physical Laboratory, England. We compare the results of fitting the BBS and BS distributions to these data.

Table 4.2 lists the MLEs of the parameters (SEs in parentheses) and the values of AIC and BIC. Based on these values, we can conclude that the BBS model gives a better fit to the data if compared to the BS model. Plots of the estimated pdfs and cdfs of the BBS and BS models fitted to these data are displayed in Figure 4.2. The plots suggest that the BBS distribution is better to the BS distribution regarding to the model fitting.

Table 4.1: MLEs of the parameters for some models fitted to the Aarset's data (the SEs are given in parentheses) and the AIC and BIC statistics.

Model	MLEs					AIC	BIC
	a	b	α	λ	γ		
BMW	0.1975	0.1647	0.0002	0.0541	1.3771	451.6	461.2
	(0.0462)	(0.0830)	(6.6931×10^{-5})	(0.0157)	(0.3387)		
BW	0.1836	0.0748	0.0007	0	2.3615	463.9	471.6
	(0.0509)	(0.0353)	(0.0004)	0	(0.1715)		
MW	1	1	0.0624	0.0233	0.3548	460.3	466.0
			(0.0267)	(0.0048)	(0.1127)		
EW	0.4668	1	0.0011	0	1.5936	480.5	486.2
	(0.0889)		(0.0010)		(0.1858)		

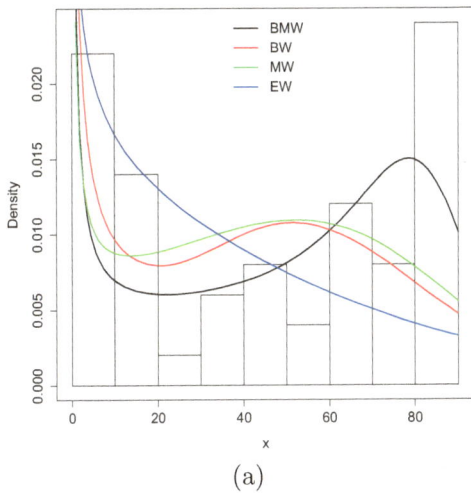

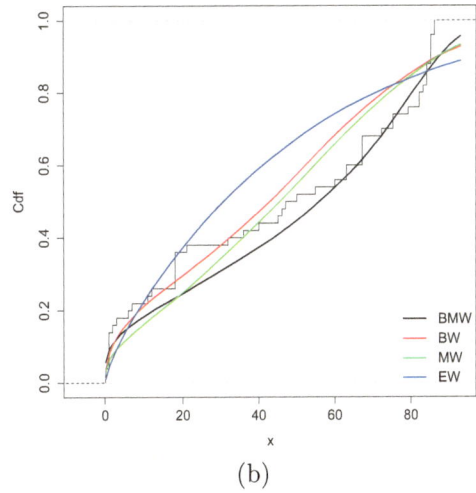

Figure 4.1: Estimated (a) pdf and (b) cdf for the BMW, BW, MW and EW models for the first data set.

Table 4.2: MLEs of the parameters for some models fitted to the glass fibres (the SEs are given in parentheses) and the AIC and BIC statistics

Model	MLEs				AIC	BIC
	a	b	α	β		
BBS	0.3638	7857.6	1.0505	30.4783	37.552	45.280
	(0.0709)	(3602.2)	(0.0101)	(0.5085)		
BS	1	1	0.2699	1.3909	48.378	52.242
			(0.0267)	(0.0521)		

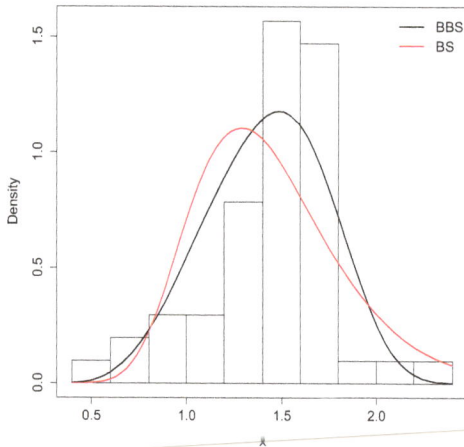 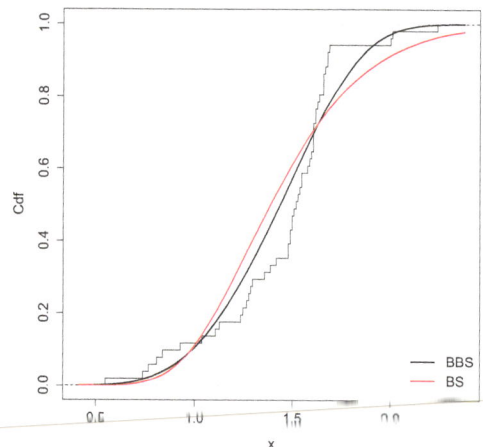

(a)   (b)

Figure 4.2: Estimated (a) pdf and (b) cdf for the BBS and BS models for the second data set.

4.7. CONCLUSIONS

One major benefit of the class of beta generalized (BG) distributions proposed by Eugene *et al.* (2002) [101] is its ability of modelling skewed data that can not be properly adjusted by existing distributions. Some BG distributions were discussed in recent literature. For example, Eugene *et al.* (2002) [101], Kotz and Nadarajah (2004) [159], Nadarajah and Gupta (2004) [160], and Nadarajah and Kotz (2006) [41] proposed the beta normal, beta Gumbel, beta Fréchet and beta exponential distributions by considering the baseline cdf $G(x)$ equal to the cdf of the normal, Gumbel, Fréchet and exponential distributions, respectively, and studied some of their properties.

In this chapter, we discuss some models in the beta family, namely: the beta generalized exponential, beta Weibull, beta Fréchet, beta modified Weibull (BMW) and beta Birnbaum–Saunders (BBS) distributions. We illustrate two applications of the beta family to real data by comparing the fits of the BMW distribution and some of its sub-models. The model parameters are estimated by maximum likelihood through the `NLMixed` subroutine in `SAS`.

More details on other beta models are given by Tahir and Nadarajah (2015) [161].

Chapter 5

The Kumaraswamy's Generalized Family of Models

Abstract:

This chapter addresses the Kumaraswamy's generalized ("Kw-G" for short) family of models. A physical motivation for the Kw-G family is presented and some of its special cases are discussed in detail. This family receives a baseline distribution as input and returns a new distribution with two additional parameters. The returned model is often more flexible than the baseline one. Several structural properties are presented and discussed for the Kw-G family. Among them, useful expansions, mgf, moments and mean deviations. Additionally, estimation and generation procedures are presented.

Keywords:: Asymptotes; Kw-G Model; Moment; Physical Motivation; Shapes.

5.1. INTRODUCTION

In life testing experiments, the data can be modeled by a wide range of distributions. Kumaraswamy (1980) [162] pioneered a distribution for double bounded random processes with applications in hydrology. In addition to the hydrological context, the Kumaraswamy (Kw) model has been adopted in related areas, such as reservoir operations and design, see, for example, Fletcher and Ponnambalam (1996) [163] and Seifi *et al.* (2000) [164].

The pdf and cdf of the Kw distribution with two shape parameters $a > 0$ and $b > 0$ in the interval $(0, 1)$ are, respectively,

$$\pi(x) = a\,b\,x^{a-1}(1 - x^a)^{b-1} \quad \text{and} \quad \Pi(x) = 1 - (1 - x^a)^b. \qquad (5.1)$$

Gauss M. Cordeiro, Rodrigo B. Silva & Abraão D. C. Nascimento

The density function (5.1) has several properties similar to those of the beta distribution but has some advantages in terms of tractability.

The Kw distribution is not widely known, although Jones (2009) [164] pointed out some differences and similarities with the beta distribution. For example, the Kw densities can be unimodal, anti-modal, increasing, decreasing or constant depending on the parameter values in a similar way of the beta distribution. He addressed several advantages of this distribution over the beta distribution such as simpler formulae for the cdf and qf and moments of the order statistics. Jones (2009) [164] also emphasized that the beta distribution has some advantages over the Kw distribution such as simpler expressions for moments and generating function and more ways for generation using physical processes.

The Kumaraswamy generalized ("Kw-G") family of distributions was proposed by Cordeiro and de Castro (2011) [87] and has the Kw distribution as the baseline model. The Kw-G family is defined as

$$F(x) = 1 - \{1 - G(x)^a\}^b, \tag{5.2}$$

Note that the two additional parameters $a > 0$ and $b > 0$ provide skewness and vary tail weights. Because of the simple form of equation (5.2), this family can be easily fitted even if the data are censored. The Kw-G family allows for greater flexibility of its tails and can be applied in several areas of engineering, medicine and biology.

Correspondingly, the pdf and hrf of this family have very simple forms:

$$f(x) = a\,b\,g(x)\,G(x)^{a-1}\left\{1 - G(x)^a\right\}^{b-1}. \tag{5.3}$$

and

$$\tau(x) = a\,b\,g(x)\,G(x)^{a-1}\{1 - G(x)^a\}^{-1}. \tag{5.4}$$

In this chapter, X denotes the random variable with density function (5.3) and we write $X \sim$ Kw-G(a, b). Each Kw-G generated distribution can be determined from a parent G distribution, which is clearly a basic exemplar of the Kw-G family when $a = b = 1$.

If $b = 1$, we obtain as a special case from (5.3) the exp-G family discussed in Chapter 2. One major benefit of the Kw-G family is to fit skewed data that can not be fitted by classic distributions. Most of the results of this chapter follow Cordeiro and de Castro (2011) [87] and Nadarajah *et al.* (2012) [88].

Based on the cdf $G(x)$ and pdf $g(x)$ of any baseline G distribution, we can associate the Kw-G density (5.3) with two extra shape parameters a and b. These parameters can generate distributions with heavier or lighter tails and control skewness and kurtosis. They can provide more flexible distributions. The Kw-G family has a wide variety of shapes and it is able to model bathtub-shaped hazard rate data. Further, it can be easily used for discriminating between the Kw-G and G distributions. If $a < 1$, then the tails of $f(x)$ will be heavier than those of $g(x)$. Similarly, if $b < 1$, then the tails of $f(x)$ will be heavier than those of $g(x)$. On the other hand, if $a > 1$, then the tails of $f(x)$ will be lighter than those of $g(x)$. Similarly, if $b > 1$, then the tails of $f(x)$ will be lighter than those of $g(x)$. The density (5.3) has an important advantage over the BG class (Eugene *et al.*, 2002 [100]) discussed in Chapter 3, since it does not involve any complicated function.

Each Kw-G distribution can be determined from a given G distribution as follows: the Kw-normal (KwN) distribution follows by taking $G(x)$ in equation (5.3) as the normal cdf. In a similar manner, the Kw-gamma (KwGa), Kw-Weibull (KwW) and Kw-Gumbel (KwGu) models follow by taking $G(x)$ to be the cdf of the gamma, Weibull and Gumbel distributions, respectively.

In this chapter, equation (5.3) is applied in some generality. The structural properties of the Kw-G family are usually simpler to derive than those of the BG family.

If $g(x)$ is a symmetric function around zero, then $f(x)$ will not be a symmetric distribution even when $a = b$. By inverting (5.2), the Kw-G qf can be expressed in terms of the baseline qf, say $Q_G(u) = G^{-1}(u)$, by $Q(u) = Q_G(\{1 - (1 - u)^{1/b}\}^{1/a})$.

5.2. PHYSICAL MOTIVATION

For a and b positive integers, a physical interpretation of the Kw-G family (5.3) can be given as follows. Suppose that a system is composed by b independent components, which in turn is composed by a independent subcomponents. Define $X_{j1}, \ldots, X_{ja}$ as the subcomponents lifetimes of the jth component (for $j = 1, \ldots, b$) with common cdf $G(x)$. Suppose that the system failure occurs if any of the b components fails and each component fails only if all a subcomponents fail. Further, denote X_j as the lifetime of the jth component (for $j = 1, \ldots, b$) and let X denote the lifetime of the entire system. Then, the cdf of X can be expressed as

$$\Pr(X \leq x) = 1 - \Pr(X_1 > x, \ldots, X_b > x) = 1 - \Pr^b(X_1 > x)$$
$$= 1 - \{1 - \Pr(X_1 \leq x)\}^b = 1 - \{1 - \Pr(X_{11} \leq x, \ldots, X_{1a} \leq x)\}^b$$
$$= 1 - \{1 - G^a(x)\}^b.$$

Hence, the Kw-G family (5.2) is precisely the time to failure distribution of the entire system.

5.3. SPECIAL Kw-G DISTRIBUTIONS

The density function (5.3) will be very tractable if $G(x)$ and $g(x)$ have simple analytic expressions. Next, we address some special Kw-G distributions. Three important special Kw-G models are explored in Chapter 6.

5.3.1. Kw-Normal (KwN)

The KwN density function comes from (5.3) by taking $G(\cdot)$ and $g(\cdot)$ to be the cdf and pdf of the normal $N(\mu, \sigma^2)$ distribution, so that

$$f(x) = \frac{ab}{\sigma}\, \phi\left(\frac{x-\mu}{\sigma}\right)\left\{\Phi\left(\frac{x-\mu}{\sigma}\right)\right\}^{a-1}\left\{1-\Phi\left(\frac{x-\mu}{\sigma}\right)^a\right\}^{b-1},$$

where $x \in \mathbb{R}$, $\mu \in \mathbb{R}$ is a location parameter, $\sigma > 0$ is a scale parameter, a and b are positive shape parameters, and $\phi(\cdot)$ and $\Phi(\cdot)$ are the pdf and cdf of the standard normal distribution, respectively. A random variable with this density is denoted by $X \sim \text{KwN}(a, b, \mu, \sigma^2)$. For $\mu = 0$ and $\sigma = 1$, we have the standard KwN distribution. Further, the KwN distribution with $a = 2$ and $b = 1$ coincides with the skew-normal distribution with shape parameter equal to one.

5.3.2. Kw-Weibull (KwW)

The Weibull cdf with parameters $\beta > 0$ and $\alpha > 0$ is $G(x) = 1 - \exp\{-(\beta x)^\alpha\}$ for $x > 0$. So, the density of the KwW(a, b, α, β) model takes the form

$$f(x) = a\,b\,\alpha\,\beta^\alpha\, x^{\alpha-1} \exp\{-(\beta x)^\alpha\}[1 - \exp\{-(\beta x)^\alpha\}]^{a-1}$$
$$\times \{1 - [1 - \exp\{-(\beta x)^\alpha\}]^a\}^{b-1}, \quad x, a, b, \alpha, \beta > 0.$$

For $\alpha = 1$, we obtain the Kw-exponential (KwE) distribution. The KwW$(1, b, 1, \beta)$ distribution becomes the exponential distribution with parameter $\beta^\star = b\beta$.

5.3.3. Kw-Gamma (KwG)

A random variable Y with shape parameter $\alpha > 0$ and scale parameter $\beta > 0$ has cdf given by

$$G_{\alpha,\beta}(x) = \frac{\gamma(\alpha, \beta x)}{\Gamma(\alpha)}, \quad x > 0.$$

A random variable X with the KwGa(a, b, α, β) density function (for $x > 0$) is

$$f(x) = \frac{a\,b\,\beta^\alpha x^{\alpha-1}\,e^{-\beta x}}{\Gamma(\alpha)^{ab}}\,\gamma(\alpha, \beta x)^{a-1}\,\left[\Gamma(\alpha) - \gamma(\alpha, \beta x)^a\right]^{b-1}.$$

For $\alpha = 1$, we obtain the Kw-exponential (KwE) distribution. Note that KwGa$(1, b, 1, \beta)$ denotes the exponential distribution with parameter $\beta^\star = b\beta$. Hence, we can have problems of identifiability for special models of the KwG family when some parameters take positive integer values.

5.3.4. Kw-Gumbel (KwGu)

The density and distribution functions of the Gumbel distribution (for $x \in \mathbb{R}$) with location parameter $\mu > 0$ and scale parameter $\sigma > 0$ are, respectively,

$$g(x) = \frac{1}{\sigma}\exp\left\{\frac{x - \mu}{\sigma} - \exp\left(\frac{x - \mu}{\sigma}\right)\right\}$$

and

$$G(x) = 1 - \exp\left\{-\exp\left(-\frac{x - \mu}{\sigma}\right)\right\}.$$

The mean and variance of the Gumbel model are $\mu - \gamma\sigma$ and $\pi^2\sigma^2/6$, respectively, where γ is the Euler's constant ($\gamma \approx 0.57722$). Inserting these expressions into (5.3) gives the KwGu(a, b, μ, σ) distribution.

5.3.5. Kw-Inverse Gaussian (KwIG)

The pdf and cdf of the inverse Gaussian distribution (for $x, \mu, \sigma > 0$) can be expressed as

$$g(x) = \frac{1}{\sqrt{2\pi\sigma^2 x^3}}\exp\left\{-\frac{(x - \mu)^2}{2\mu^2\sigma^2 x}\right\}$$

and

$$G(x) = \Phi\left(\frac{1}{\sqrt{\sigma^2 x}}\left[\frac{x}{\mu} - 1\right]\right) + \exp\left(\frac{2}{\mu\sigma^2}\right)\Phi\left(-\frac{1}{\sqrt{\sigma^2 x}}\left[\frac{x}{\mu} + 1\right]\right).$$

The expectation and variance are equal to μ and $\sigma^2\mu^3$, respectively. Inserting these expressions into (5.3) gives the KwIG(a, b, μ, σ^2) distribution.

5.3.6. Kw-Chen (KwChen)

The KwChen density function (for $x > 0$) is given by

$$f(x) = a\,b\,\lambda\,\beta\,x^{\beta-1}\exp(x^\beta) \times$$
$$\exp\left\{\lambda[1 - \exp(x^\beta)]\right\}\left\{1 - \exp\{\lambda[1 - \exp(x^\beta)]\}\right\}^{a-1} \times$$
$$\left\{1 - \left(1 - \exp\{\lambda[1 - \exp(x^\beta)]\}\right)^a\right\}^{b-1}, \tag{5.5}$$

where $\lambda > 0$ and $\beta > 0$. We write $X \sim\text{KwChen}(a, b, \lambda, \beta)$ if X is a random variable with density function (5.5). It becomes the Chen distribution (Chen, 2000 [165]) when $a = b = 1$. The KwChen survival function is

$$S(x) = [1 - (1 - \exp\{\lambda[1 - \exp(x^\beta)]\})^a]^b.$$

5.3.7. Kw-XTG (KwXTG)

The KwXTG density function (for $x > 0$) is

$$f(x) = a\,b\,\lambda\,\beta\left(\frac{x}{\alpha}\right)^{\beta-1}\exp\left\{\left(\frac{x}{\alpha}\right)^\beta + \lambda\,\alpha\left[1 - \exp\left(\left(\frac{x}{\alpha}\right)^\beta\right)\right]\right\} \times$$
$$\left[1 - \exp\left\{\lambda\,\alpha\left[1 - \exp\left(\left(\frac{x}{\alpha}\right)^\beta\right)\right]\right\}\right]^{a-1} \times$$
$$\left\{1 - \left(1 - \exp\left\{\lambda\,\alpha\left[1 - \exp\left(\left(\frac{x}{\alpha}\right)^\beta\right)\right]\right\}\right)^a\right\}^{b-1}, \tag{5.6}$$

where $\lambda > 0$, $\alpha > 0$ and $\beta > 0$. We write $X \sim\text{KwXTG}(a, b, \lambda, \alpha, \beta)$ if X is a random variable with density (5.6). For $a = b = 1$, it reduces to the XTG distribution (Xie *et al.*, 2002 [156]). The KwXTG survival function is

$$S(x) = \left[1 - \left(1 - \exp\left\{\lambda\,\alpha\left[1 - \exp\left(\left(\frac{x}{\alpha}\right)^\beta\right)\right]\right\}\right)^a\right]^b.$$

5.3.8. Kw-Flexible Weibull (KwFW)

The KwFW density function (for $x > 0$) has the form

$$f(x) = a\,b\left(\alpha + \frac{\beta}{x^2}\right)\exp\left(\alpha\,x - \frac{\beta}{x}\right)\exp\left\{-\exp\left(\alpha\,x - \frac{\beta}{x}\right)\right\} \times$$
$$\left[1 - \exp\left\{-\exp\left(\alpha\,x - \frac{\beta}{x}\right)\right\}\right]^{a-1} \times$$
$$\left\{1 - \left(1 - \exp\left\{-\exp\left(\alpha\,x - \frac{\beta}{x}\right)\right\}\right)^a\right\}^{b-1}, \tag{5.7}$$

where $\alpha > 0$ and $\beta > 0$. We write $X \sim \mathrm{KwFW}(a, b, \alpha, \beta)$ if X is a random variable with density function (5.7). For $a = b = 1$, it reduces to the flexible Weibull (FW) model (Bebbington *et al.*, 2007 [166]). The corresponding survival function is

$$S(x) = \left[1 - \left(1 - \exp\left\{-\exp\left(\alpha x - \frac{\beta}{x}\right)\right\}\right)^a\right]^b.$$

5.4. ASYMPTOTES AND SHAPES

The asymptotes of (5.2), (5.3) and (5.4) as $G(x) \to 0$ and $x \to \infty$ are

$$F(x) \sim aG^a(x) \text{ as } G(x) \to 0, \ 1 - F(x) \sim \{1 - G^a(x)\}^b \text{ as } x \to \infty,$$

$$f(x) \sim abg(x)G^{a-1}(x) \quad \text{as} \quad G(x) \to 0, \quad f(x) \sim abg(x)\{1 - G^a(x)\}^{b-1}$$

as $x \to \infty$,

$$\tau(x) \sim abg(x)G^{a-1}(x) \quad \text{as} \quad G(x) \to 0, \quad \tau(x) \sim \frac{abg(x)}{1 - G^a(x)} \text{ as } x \to \infty.$$

The shapes of the pdf (5.3) and the hrf (5.4) can be described analytically. The critical points of the pdf are the roots of the equation:

$$\frac{g'(x)}{g(x)} + (a - 1)\frac{g(x)}{G(x)} = a(b - 1)\frac{g(x)G^{a-1}(x)}{1 - G^a(x)}. \tag{5.8}$$

There may be more than one root to (5.8). Let $\lambda(x) = d^2 \log[f(x)]/dx^2$. If $x = x_0$ is a root of (5.8) then it corresponds to a local maximum, a local minimum or a point of inflexion depending on whether $\lambda(x_0) < 0$, $\lambda(x_0) > 0$ or $\lambda(x_0) = 0$, where

$$\begin{aligned}
\lambda(x) = & \frac{g(x)g''(x) - g'(x)^2}{g^2(x)} + (a - 1)\frac{G(x)g'(x) - g^2(x)}{G^2(x)} \\
& -a(b-1)\frac{G^{a-2}(x)\{(a-1)g^2(x) + G(x)g'(x)\}}{1 - G^a(x)} \\
& -a^2(b-1)\frac{G^{2a-2}(x)g^2(x)}{\{1 - G^a(x)\}^2}.
\end{aligned}$$

The critical points of the hrf of X are the roots of the equation:

$$\frac{g'(x)}{g(x)} + (a - 1)\frac{g(x)}{G(x)} = -a\frac{g(x)G^{a-1}(x)}{1 - G^a(x)}. \tag{5.9}$$

There may be more than one root to (5.9). Let $\delta(x) = d^2 \log[h(x)]/dx^2$. If $x = x_0$ is a root of (5.9), then it corresponds to a local maximum, a local minimum or a point of inflexion depending on whether $\delta(x_0) < 0$, $\delta(x_0) > 0$ or $\delta(x_0) = 0$, where

$$\delta(x) = \frac{g(x)\,g''(x) - g'(x)^2}{g^2(x)} + (a-1)\frac{G(x)g'(x) - g^2(x)}{G^2(x)}$$
$$-a\frac{G^{a-2}(x)\left\{(a-1)g^2(x) + G(x)g'(x)\right\}}{1 - G^a(x)} - a^2\frac{G^{2a-2}(x)g^2(x)}{\{1 - G^a(x)\}^2}.$$

5.5. SIMULATION

A simple method for simulating the Kw-G distribution uses the inversion method. The qf corresponding to (5.2) is directly obtained from the qf associated with $G(x)$ by

$$F^{-1}(x) = Q_G\left\{\left[1 - (1-x)^{1/b}\right]^{1/a}\right\}, \tag{5.10}$$

where $Q_G(\cdot) = G^{-1}(\cdot)$. So, one can generate Kw-G variates by

$$X = G^{-1}\left\{\left[1 - (1-U)^{1/b}\right]^{1/a}\right\},$$

where U is a uniform variate on the unit interval $[0,1]$.

5.6. USEFUL EXPANSIONS

We derive two useful expansions for (5.23) and (5.3) based on the exp-G distribution. For an arbitrary baseline cdf $G(x)$, the pdf and cdf of a random variable Z having the exp-G distribution with power parameter $a > 0$ (see Section 2.1) are

$$h_a(x) = a\,g(x)G^{a-1}(x) \qquad \text{and} \qquad H_a(x) = G^a(x), \tag{5.11}$$

respectively.

Expanding the binomial terms in (5.2) and (5.3), the cdf and pdf of X can be expressed as

$$F(x) = 1 - \sum_{i=0}^{\infty}(-1)^i \binom{b}{i} H_{i\,a}(x), \tag{5.12}$$

and

$$f(x) = a^{-1} \sum_{i=0}^{\infty} \frac{w_i}{(i+1)} h_{(i+1)\,a}(x) = g(x) \sum_{i=0}^{\infty} w_i \, G(x)^{(i+1)\,a-1}, \qquad (5.13)$$

where $w_i = (-1)^k \, a\, b \binom{b-1}{i}$, $h_{(i+1)\,a}(x)$ and $H_{i\,a}(x)$ are the pdf and cdf of the exp-G$((i+1)\,a)$ and exp-G(ia) distributions, respectively. Then, some mathematical properties of the Kw-G model can follow from (5.13) by knowing those corresponding exp-G properties.

5.7. MOMENTS

A first representation for the nth moment of X follows from (5.13) as

$$\mu'_n = \mathrm{E}(X^n) = a^{-1} \sum_{i=0}^{\infty} \frac{w_i}{(i+1)} \, \mathrm{E}(Y_i^n),$$

where $Y_i \sim$ exp-G$((i+1)\,a)$. Expressions for moments of some exp-G distributions are given by Nadarajah and Kotz (2006) [41], which can be used to produce $\mathrm{E}(X^n)$.

A second representation for $\mathrm{E}(X^n)$ follows from (5.13) as

$$\mu'_n = \mathrm{E}(X^n) = \sum_{i=0}^{\infty} w_i \, \tau(n, (i+1)a - 1), \qquad (5.14)$$

where $\tau(n, a)$ can be expressed in terms of the baseline qf $Q_G(u)$ as

$$\tau(n, a) = \int_{-\infty}^{\infty} x^n \, G(x)^a \, g(x)\mathrm{d}x = \int_0^1 Q_G(u)^n \, u^a \mathrm{d}u. \qquad (5.15)$$

The ordinary moments for most Kw-G distributions can be determined directly from (5.14) and (5.15) at least numerically. Now, we provide two examples. The moments of the KwE (with parameter $\lambda > 0$) distribution are given by

$$\mathrm{E}(X^n) = n!\, \lambda^n \sum_{i,j=0}^{\infty} \frac{(-1)^{n+j}\, w_i}{(j+1)^{n+1}} \binom{(i+1)a - 1}{j}.$$

For the Kw-standard logistic (KwSL) distribution, where $G(x) = (1 + \mathrm{e}^{-x})^{-1}$, we obtain from Prudnikov *et al.* (1986, Section 2.6.13, equation 4 [145]) (for $t < 1$)

$$\mathrm{E}(X^n) = \sum_{i=0}^{\infty} w_i \left(\frac{\partial}{\partial t} \right)^n B\left(t + (i+1)a, 1 - t \right)\Big|_{t=0}.$$

The central moments (μ_s) and cumulants (κ_s) of X can be determined from the ordinary moments as given in Section 1.4. The skewness $\gamma_1 = \kappa_3/\kappa_2^{3/2}$ and kurtosis $\gamma_2 = \kappa_4/\kappa_2^2$ are evaluated from the third and fourth standardized cumulants.

For some G distributions, Cordeiro and Nadarajah (2011) [82] derived closed-form expressions for the moments of X as linear functions of the PWMs of the baseline random variable Y. These moments can depend on special mathematical functions.

5.8. GENERATING FUNCTION

We give three formulae for the mgf $M(t) = E(\mathrm{e}^{tX})$ of X. The first one is given by

$$
\begin{aligned}
M(t) &= a\,b\,E\left[\mathrm{e}^{tX}\,G^{a-1}(X)\,\{1 - G^a(X)\}^{b-1}\right] \\
&= a\,b\sum_{k=0}^{\infty}(-1)^k\binom{b-1}{k}E\left[\mathrm{e}^{tX}\,U^{[(k+1)a-1]}\right],
\end{aligned}
$$

where U is a uniform random variable on the unit interval. However, note that X and U are not independent.

A second formula for $M(t)$ is obtained from (5.13)

$$
M(t) = a^{-1}\sum_{i=0}^{\infty}\frac{w_i}{(i+1)}\,M_i(t),
$$

where $M_i(t)$ is the mgf of $Y_i \sim \exp\text{-G}((i+1)\,a)$. Then, for several Kw-G distributions, $M(t)$ can be determined from the mgfs of the G distributions.

A third formula for $M(t)$ can be derived from (5.13) as

$$
M(t) = \sum_{i=0}^{\infty} w_i\,\rho(t, a(i+1) - 1), \tag{5.16}
$$

where

$$
\rho(t, a) = \int_{-\infty}^{\infty}\mathrm{e}^{tx}\,G(x)^a\,g(x)\mathrm{d}x = \int_{0}^{1}\exp\{t\,Q(u)\}\,u^a\,\mathrm{d}u. \tag{5.17}
$$

We can obtain the mgfs of various Kw-G distributions from (5.16) and (5.17). For example, for the KwE (with real parameter $\lambda > 0$ and $t < \lambda^{-1}$) and KwSL (for $t < 1$) distributions, we have

$$M(t) = \sum_{i=0}^{\infty} w_i \, B\left((i+1)a, 1 - \lambda t\right)$$

and

$$M(t) = \sum_{i=0}^{\infty} w_i \, B\left(t + (i+1)a, 1 - t\right),$$

respectively.

5.9.   MEAN DEVIATIONS

Let $X \sim \text{KwG}(a, b)$. The mean deviations about the mean $(\delta_1 = E(|X - \mu_1'|))$ and about the median $(\delta_2 = E(|X - M|))$ can be expressed as

$$\delta_1 = 2\mu_1' \, F(\mu_1') - 2m_1(\mu_1') \quad \text{and} \quad \delta_2 = \mu_1' - 2m_1(M), \qquad (5.18)$$

respectively, where $\mu_1' = E(X)$, $M = Median(X)$ denotes the median, $F(\mu_1')$ comes from (5.2) and $m_1(z) = \int_{-\infty}^{z} x \, f(x) \, dx$ denotes the first incomplete moment of X.

The median M follows from (5.10) as

$$M = Q_G \left(\left[1 - 2^{-1/b}\right]^{1/a} \right).$$

Setting $u = G(x)$ in equation (5.13) gives

$$m_1(z) = \sum_{k=0}^{\infty} w_k \, T_k(z), \qquad (5.19)$$

where $T_k(z)$ can be expressed in terms of $Q_G(u)$ as

$$T_k(z) = \int_0^{G(z)} Q_G(u) \, u^{(k+1)a-1} \, du. \qquad (5.20)$$

The quantity $m_1(z)$ is useful to plot Lorenz and Bonferroni curves in some areas (economics, reliability, demography, insurance and medicine, among others). For a given probability π, they are defined by $L(\pi) = T_1(q)/\mu_1'$ and $B(\pi) = T_1(q)/(\pi \, \mu_1')$, respectively, where $\mu_1' = E(X)$ and $q = F^{-1}(\pi) = Q_G\{[1 - (1 - \pi)^{1/b}]^{1/a}\}$ can be evaluated from the baseline qf.

The mean deviations of any Kw-G distribution can be computed from equations (5.18)-(5.20). For example, the quantities $T_k(z)$ for the KwE (with parameter

λ), KwSL and Kw-Pareto (KwPa) $(G(x) = 1 - (1 + x)^{-\nu}$ with parameter $\nu > 0)$ distributions are found using the generalized binomial expansion as

$$T_k(z) = \lambda^{-1}\,\Gamma((k+1)a+1) \sum_{j=0}^{\infty} \frac{(-1)^j\,\{1 - \exp(-j\lambda z)\}}{\Gamma((k+1)a+1-j)\,(j+1)!},$$

$$T_k(z) = \frac{1}{\Gamma(k)} \sum_{j=0}^{\infty} \frac{(-1)^j\,\Gamma((k+1)a+j)\,\{1 - \exp(-jz)\}}{(j+1)!}$$

and

$$T_k(z) = \sum_{j=0}^{\infty} \sum_{r=0}^{j} \frac{(-1)^j \binom{(k+1)a}{j} \binom{j}{r}}{(1-r\nu)}\, z^{1-r\nu},$$

respectively.

An alternative representation for $m_1(z)$ can be derived from (5.13) as

$$m_1(z) = \int_{-\infty}^{z} x\, f(x)\, dx = a^{-1} \sum_{k=0}^{\infty} \frac{w_k}{(k+1)}\, J_k(z), \qquad (5.21)$$

where

$$J_k(z) = \int_{-\infty}^{z} x\, h_{(k+1)a}(x)\, \mathrm{d}x. \qquad (5.22)$$

The integral (5.22) is the main quantity to determine the mean deviations of the exp-G distributions. Hence, the Kw-G mean deviations depend only on the exp-G mean deviations.

So, alternative representations for δ_1 and δ_2 are given by

$$\delta_1 = 2\mu_1' F(\mu_1') - 2a^{-1} \sum_{k=0}^{\infty} \frac{w_k}{k+1}\, J_k(\mu_1'), \quad \delta_2 = \mu_1' - 2a^{-1} \sum_{k=0}^{\infty} \frac{w_k}{k+1}\, J_k(M).$$

Next, a simple application is provided for the KwW distribution. The EW distribution with power parameter $(k+1)a$ has density function (for $x > 0$) given by

$$h_{(k+1)a}(x) = (k+1)\, a\, c\, \beta^c\, x^{c-1} \exp\{-(\beta x)^c\}\,[1 - \exp\{-(\beta x)^c\}]^{(k+1)a-1}$$

and then

$$J_k(z) = (k+1)\, a\, c\, \beta^c \int_0^z x^c \exp\{-(\beta x)^c\}[1 - \exp\{-(\beta x)^c\}]^{(k+1)a-1} dx$$

$$= (k+1)\, a\, c\, \beta^c \sum_{r=0}^{\infty}(-1)^r \binom{(k+1)a-1}{r} \int_0^z x^c \exp\{-(r+1)(\beta x)^c\} dx.$$

We can evaluate the last integral using the incomplete gamma function as

$$J_k(z) = (k+1)\, a\, \beta^{-1} \sum_{r=0}^{\infty} \frac{(-1)^r \binom{(k+1)a-1}{r}}{(r+1)^{1+c^{-1}}}\, \gamma(1 + c^{-1}, (r+1)(\beta z)^c).$$

5.10.   RELATION WITH THE BETA-G

Consider starting from the baseline cdf $G(x)$ and pdf $g(x)$, Eugene *et al.*(2002) [100] defined the beta-G(a, b) density function (see Chapter 3) by

$$f(x) = \frac{1}{B(a,b)} G(x)^{a-1} \{1 - G(x)\}^{b-1} g(x), \qquad (5.23)$$

where $a > 0$ and $b > 0$ are two additional parameters to those parameters of G. We can note that if $Z \sim$exp-G(a) with cdf $H_a(x) = G(x)^a$, then the beta-$H_a(1, b)$ distribution is identical to the Kw-G(a, b) distribution. So, the beta-G$(1, b)$ generator applied to the exp-G(a) model gives the Kw-G(a, b) distribution. Some properties of special Kw-G models can be derived in this way. Clearly, if Z has the beta-G$(1, b)$ distribution, then $X = G^{-1}(Z^{1/a})$ will have the Kw-G(a, b) distribution.

Next, we use this result to obtain some properties of the KwE(a, b, λ) model from the beta exponentiated exponential (BEE) (Barreto-Souza *et al.*, 2010 [167]) distribution. Consider the exponential distribution with parameter λ. Its properties can be derived from those of the three-parameter BEE$(1, b, \lambda, a)$ model. The KwE density function with parameters a, b and λ is given by

$$f(x) = a\, b\, \lambda\, e^{-\lambda x}(1 - e^{-\lambda x})^{a-1} \{1 - (1 - e^{-\lambda x})^a\}^{b-1}, \quad x > 0.$$

For $b > 0$ real non-integer, the moments of X are

$$\mu'_r = E(X^r) = \frac{a\Gamma(b+1)}{\lambda^r} \sum_{j=0}^{\infty} (-1)^{j+r} \binom{b-1}{j} \frac{\partial^r B(p, (j+1)a)}{\partial p^r}\bigg|_{p=1}.$$

For $t < \lambda$, the KwE mgf follows immediately from Barreto-Souza *et al.* (2010) [167] (for $b > 0$ real non-integer) as

$$M(t) = a\, b \sum_{j=0}^{\infty} (-1)^j \binom{b-1}{j} B(1 - t/\lambda, (j+1)a).$$

5.11.   ESTIMATION

We estimate the parameters of the Kw-G family for uncensored data by maximum likelihood.

Let $x_1, \ldots, x_n$ be an observed sample from the Kw-G family with pdf (5.3). Suppose that $g(x)$ is parameterized by a vector $\boldsymbol{\theta}$ of length p. The log likelihood

(LL) function for the parameters $\boldsymbol{\tau} = (a, b, \boldsymbol{\theta})$ is

$$\ell(a, b, \boldsymbol{\theta}) = n \log a + n \log b + \sum_{j=1}^{n} \log g\left(x_j; \boldsymbol{\theta}\right) + (a - 1) \sum_{j=1}^{n} \log G\left(x_j; \boldsymbol{\theta}\right)$$

$$+ (b - 1) \sum_{j=1}^{n} \log \left\{1 - G^a\left(x_j; \boldsymbol{\theta}\right)\right\}. \tag{5.24}$$

The MLEs are the simultaneous solutions of the equations:

$$\frac{n}{a} + \sum_{j=1}^{n} \log G\left(x_j; \boldsymbol{\theta}\right) - (b - 1) \sum_{j=1}^{n} \frac{G^a\left(x_j; \boldsymbol{\theta}\right) \log G\left(x_j; \boldsymbol{\theta}\right)}{1 - G^a\left(x_j; \boldsymbol{\theta}\right)} = 0,$$

$$\frac{n}{b} + \sum_{j=1}^{n} \log \left\{1 - G^a\left(x_j; \boldsymbol{\theta}\right)\right\} = 0$$

and

$$\sum_{j=1}^{n} \frac{1}{g\left(x_j; \boldsymbol{\theta}\right)} \frac{\partial g\left(x_j; \boldsymbol{\theta}\right)}{\partial \theta_k} + (a - 1) \sum_{j=1}^{n} \frac{1}{G\left(x_j; \boldsymbol{\theta}\right)} \frac{\partial G\left(x_j; \boldsymbol{\theta}\right)}{\partial \theta_k}$$

$$- a(b - 1) \sum_{j=1}^{n} \frac{G^{a-1}\left(x_j; \boldsymbol{\theta}\right)}{1 - G^a\left(x_j; \boldsymbol{\theta}\right)} \frac{\partial G\left(x_j; \boldsymbol{\theta}\right)}{\partial \theta_k} = 0.$$

For interval estimation of the parameters in $\boldsymbol{\tau}$ and tests of hypotheses, one requires the expected information matrix, say K. Interval estimation for the model parameters can be performed using standard likelihood theory. The elements of this matrix for (5.24) can be obtained from Nadarajah *et al.* (2013) [26].

We evaluate the maximized unrestricted and restricted log-likelihoods to perform LR tests for special models of the Kw-G distribution (see Section 1.2.3). There are different forms of censored data in lifetime analysis: type I censoring, type II censoring, *etc.* We consider multi-censored general data, where there are n subjects such that:

- n_0 are known to have failed at the times $x_1, \ldots, x_{n_0}$.

- n_1 are known to have failed in the interval $[s_{j-1}, s_j]$, $j = 1, \ldots, n_1$.

- n_2 survived to a time r_j, $j = 1, \ldots, n_2$, but not observed any longer.

Note that $n = n_0 + n_1 + n_2$.

The type I censoring and type II censoring are special cases of multicensoring. The log-likelihood function for the parameters, $(a, b, \boldsymbol{\theta})$, for these data is

$$\ell(a, b, \boldsymbol{\theta}) = n_0 \log a + n_0 \log b + \sum_{j=1}^{n_0} \log g\left(x_j; \boldsymbol{\theta}\right)$$

$$+ (a - 1) \sum_{j=1}^{n_0} \log G\left(x_j; \boldsymbol{\theta}\right) + (b - 1) \sum_{j=1}^{n_0} \log \left\{1 - G^a\left(x_j; \boldsymbol{\theta}\right)\right\}$$

$$+ \sum_{j=1}^{n_1} \log \left\{\left[1 - G^a\left(s_{j-1}; \boldsymbol{\theta}\right)\right]^b - \left[1 - G^a\left(s_j; \boldsymbol{\theta}\right)\right]^b\right\}$$

$$+ b \sum_{j=1}^{n_2} \log \left\{1 - G^a\left(r_j; \boldsymbol{\theta}\right)\right\}. \tag{5.25}$$

Then, the MLEs are the simultaneous solutions of the equations:

$$\frac{n_0}{a} + \sum_{j=1}^{n_0} \log G\left(x_j; \boldsymbol{\theta}\right) - (b - 1) \sum_{j=1}^{n_0} \frac{G^a\left(x_j; \boldsymbol{\theta}\right) \log G\left(x_j; \boldsymbol{\theta}\right)}{1 - G^a\left(x_j; \boldsymbol{\theta}\right)}$$

$$+ b \sum_{j=1}^{n_1} \frac{U\left(s_j; \boldsymbol{\theta}\right) - U\left(s_{j-1}; \boldsymbol{\theta}\right)}{\left[1 - G^a\left(s_{j-1}; \boldsymbol{\theta}\right)\right]^b - \left[1 - G^a\left(s_j; \boldsymbol{\theta}\right)\right]^b}$$

$$- b \sum_{j=1}^{n_2} \frac{G^a\left(r_j; \boldsymbol{\theta}\right) \log G\left(r_j; \boldsymbol{\theta}\right)}{1 - G^a\left(r_j; \boldsymbol{\theta}\right)} = 0,$$

$$\frac{n_0}{b} + \sum_{j=1}^{n_0} \log \left\{1 - G^a\left(x_j; \boldsymbol{\theta}\right)\right\} - \sum_{j=1}^{n_1} \frac{V\left(s_j; \boldsymbol{\theta}\right) - V\left(s_{j-1}; \boldsymbol{\theta}\right)}{\left[1 - G^a\left(s_{j-1}; \boldsymbol{\theta}\right)\right]^b - \left[1 - G^a\left(s_j; \boldsymbol{\theta}\right)\right]^b}$$

$$+ \sum_{j=1}^{n_2} \log \left\{1 - G^a\left(r_j; \boldsymbol{\theta}\right)\right\} = 0$$

and

$$\sum_{j=1}^{n_0} \frac{1}{g\left(x_j; \boldsymbol{\theta}\right)} \frac{\partial g\left(x_j; \boldsymbol{\theta}\right)}{\partial \theta_k} + (a-1)\sum_{j=1}^{n_0} \frac{1}{G\left(x_j; \boldsymbol{\theta}\right)} \frac{\partial G\left(x_j; \boldsymbol{\theta}\right)}{\partial \theta_k}$$

$$- a(b-1)\sum_{j=1}^{n_0} \frac{G^{a-1}\left(x_j; \boldsymbol{\theta}\right)}{1-G^a\left(x_j; \boldsymbol{\theta}\right)} \frac{\partial G\left(x_j; \boldsymbol{\theta}\right)}{\partial \theta_k}$$

$$+ ab\sum_{j=1}^{n_1} \frac{W\left(s_j; \boldsymbol{\theta}\right) - W\left(s_{j-1}; \boldsymbol{\theta}\right)}{\left[1-G^a\left(s_{j-1}; \boldsymbol{\theta}\right)\right]^b - \left[1-G^a\left(s_j; \boldsymbol{\theta}\right)\right]^b}$$

$$- ab\sum_{j=1}^{n_2} \frac{G^{a-1}\left(r_j; \boldsymbol{\theta}\right)\partial G\left(r_j; \boldsymbol{\theta}\right)/\partial \theta_k}{1-G^a\left(r_j; \boldsymbol{\theta}\right)} = 0,$$

where $U(s; \boldsymbol{\theta}) = \{1-G^a(s)\}^{b-1}G^a(s)\log[G(s)]$, $V(s; \boldsymbol{\theta}) = \{1-G^a(s)\}^b \log\{1-G^a(s)\}$ and $W(s; \boldsymbol{\theta}) = \{1-G^a(s)\}^{b-1}G^{a-1}(s)\partial G(s)/\partial \theta_k$. The observed information matrix corresponding to (5.25) is too complicated to be presented here.

5.12.   CONCLUSIONS

Cordeiro and de Castro (2010) [87] pioneered the Kumaraswamy-G ("Kw-G") family to extend many well-known distributions in the literature. Nadarajah *et al.* (2012) [168] demonstrated that the family density function can be given as a linear combination of exp-G densities.

In this chapter, we explore their results and present eight special models. We derive general explicit expressions for some of its mathematical quantities including asymptotes and shapes, simulation, explicit formulae for the ordinary and incomplete moments, quantile and generating functions and mean deviations, among others.

These expressions are simple to be implemented in analytic and numerical statistical platforms. We adopt the maximum likelihood method to estimate the unknown parameters of the family including the case of censoring data.

Chapter 6

Special Kumaraswamy Generalized Models

Abstract: In this chapter, three special cases of the Kumaraswamy generalized family of distributions are presented. Some mathematical characteristics are provided such as the moments and generating function. An expanded expression for the density function and some special cases are presented. For illustrative purposes, practical examples of the KwG models are reported by means of applications to empirical data.

Keywords: Kw-G model; KwBXII; KwGum; KwW; Moment.

The cdf and pdf of the Kumaraswamy-G (Kw-G) family (Cordeiro and de Castro, 2010 [87]) are given by

$$F(x) = 1 - \{1 - G(x)^a\}^b \qquad (6.1)$$

and

$$f(x) = a\,b\,g(x)\,G(x)^{a-1}\left\{1 - G(x)^a\right\}^{b-1}, \qquad (6.2)$$

respectively, for $a > 0$ and $b > 0$, where $G(x)$ and $g(x)$ are the cdf and pdf of an arbitrary baseline distribution, respectively. We denote by $X \sim \text{Kw-G}(a, b)$ a random variable with cdf (6.1) and pdf (6.2).

Next, we present some mathematical characteristics of three important special models of the Kw-G family, namely: the Kumaraswamy Weibull (Cordeiro *et al.*, 2010 [84]), Kumaraswamy Burr XII (Paranaíba *et al.*, 2011 [117] and 2013 [169]) and Kumaraswamy Gumbel (Cordeiro *et al.*, 2012d [170]) distributions.

Gauss M. Cordeiro, Rodrigo B. Silva & Abraão D. C. Nascimento

6.1. KUMARASWAMY WEIBULL

The Weibull distribution has been widely used along several decades in a variety of research areas, such as engineering, medicine and biological sciences, among others. In the last two decades, several extensions of the Weibull distribution were proposed. In this sense, we have the exponentiated Weibull (EW) (Mudholkar *et al.*, 1995 [36], Mudholkar and Hutson 1996 [59]), additive Weibull (Xie and Lai, 1995 [171]), extended Weibull (Xie *et al.*, 2002 [156]), modified Weibull (MW) (Lai *et al.*, 2003 [72]), beta exponential (BE) (Nadarajah and Kotz, 2005 [41]), beta Weibull (BW) (Lee *et al.*, 2007 [81]), extended flexible Weibull (Bebbington *et al.*, 2007 [166]), generalized modified Weibull (GMW) (Carrasco *et al.*, 2008 [5]) and generalized inverse Weibull (Gusmão *et al.*, 2009 [140]) distributions.

In this section, we address some mathematical properties of the KwW model in order to increase the range of possible applications. Note that the Weibull distribution is a basic exemplar of the KwW distribution. Most of the KwW properties presented here were derived by Cordeiro *et al.* (2010a) [84].

The cdf and pdf of the KwW distribution are obtained from (6.1) and (6.2) by taking $G_{\alpha,\beta}(x) = 1 - \exp\{-(\beta x)^\alpha\}$, *i.e.*, the Weibull cdf with parameters $\alpha > 0$ and $\beta > 0$. Hence, we obtain

$$F(x) = 1 - \{1 - [1 - \exp\{-(\beta x)^\alpha\}]^a\}^b \tag{6.3}$$

and

$$\begin{aligned} f(x) &= a\,b\,\alpha\,\beta^\alpha\,x^{\alpha-1}\exp\{-(\beta x)^\alpha\}[1 - \exp\{-(\beta x)^\alpha\}]^{a-1} \\ &\quad \times \{1 - [1 - \exp\{-(\beta x)^\alpha\}]^a\}^{b-1}. \end{aligned} \tag{6.4}$$

Hereafter, the random variable X following (6.4) with parameters a, b, α and β is denoted by $X \sim \mathrm{KwW}(a, b, \alpha, \beta)$.

It is clear that the Weibull, EW and EE models are special cases of (6.4) when $a = b = 1$, $b = 1$, and $\alpha = b = 1$, respectively. The KwW distribution (6.4) is much more flexible than its special cases.

The hrf of X is

$$\tau(x) = \frac{ab\alpha\,\beta^\alpha\,x^{\alpha-1}\exp\{-(\beta x)^\alpha\}\,[1 - \exp\{-(\beta x)^\alpha\}]^{a-1}}{1 - [1 - \exp\{-(\beta x)^\alpha\}]^a}. \tag{6.5}$$

Further, the asymptotes of $f(x)$ and $F(x)$ as $x \to 0, \infty$ are given by

$$f(x) \sim a\,b\,\alpha\beta^{a\alpha}x^{a\alpha-1} \quad \text{as} \quad x \to 0,$$

$$f(x) \sim a^b b\,\alpha\,\beta^\alpha\,x^{\alpha-1}\exp\left\{-b(\beta x)^\alpha\right\} \quad \text{as} \quad x \to \infty,$$

$$F(x) \sim b(\beta x)^{a\alpha} \quad \text{as} \quad x \to 0,$$

$$1 - F(x) \sim a^b \exp\left\{-b(\beta x)^\alpha\right\} \quad \text{as} \quad x \to \infty.$$

Note that the upper tail of $f(x)$ is exponential and the lower tail is polynomial.

6.1.1. Linear Representation

Expanding the binomial $\{1 - G^a(x)\}^{b-1}$ in equation (6.2), the Kw-G family density can be expressed as

$$f(x) = \sum_{j=0}^{\infty} \frac{(-1)^j b}{(j+1)} \binom{b-1}{j} h_{(j+1)a}(x), \tag{6.6}$$

where $h_a(x) = ag(x) G(x)^{a-1}$ represents the exp-G density with parameter $a > 0$ (Eugene *et al.*, 2002 [100]) (see Section 2.1). In Chapter 2, we obtain some mathematical properties of the exponentiated models.

The KwW density can be expressed as a linear combination of Weibull densities by applying (6.6) to the Weibull distribution and expanding the generalized binomial. We obtain

$$f(x) = \sum_{k=0}^{\infty} w_k \, g_{\alpha,\beta_k}(x), \tag{6.7}$$

where $g_{\alpha,\beta_k}(x)$ is the Weibull density function with parameters α and $\beta_k = \beta (k+1)^{1/\alpha}$, and the coefficients w_k are given by

$$w_k = \sum_{j=0}^{\infty} \frac{a \, b \, (-1)^{j+k}}{(k+1)} \binom{b-1}{j} \binom{(j+1) \, a - 1}{k}. \tag{6.8}$$

It is easily verified using `Maple` that $\sum_{k=0}^{\infty} w_k = 1$ as expected. By integrating (6.7), we have

$$F(x) = \sum_{k=0}^{\infty} w_k \, G_{\alpha,\beta_k}(x). \tag{6.9}$$

6.1.2. Moments

Based on equation (6.7), some structural properties like ordinary, incomplete, factorial and inverse moments of X can be determined as infinite linear combinations of the corresponding Weibull quantities. For example, the sth ordinary moment of the Weibull distribution with parameters α and β is $\delta'_s = \beta^{-s}\,\Gamma(s/\alpha + 1)$. Then, the sth generalized moment of X reduces to

$$\mu'_s = E(X^s) = \beta^{-s}\,\Gamma(s/\alpha + 1) \sum_{k=0}^{\infty} \frac{w_k}{(k+1)^{s/\alpha}}. \tag{6.10}$$

For $a = b = 1$, the sth moment of the Weibull distribution is given exactly by equation (6.10). By using well-known results, the skewness and kurtosis measures can be easily obtained from the ordinary moments.

6.1.3. Generating Function

We provide two representations for the mgf $M(t) = E(e^{tX})$ of X based on equation (6.7). First, we can write

$$M(t) = \sum_{k=0}^{\infty} v_k\, I_k(t), \tag{6.11}$$

where $\beta_k = \beta(k+1)^{1/\alpha}$ (as given in the previous section), $v_k = (k+1)\,\alpha\beta^\alpha w_k$, and

$$I_k(t) = \int_0^\infty x^{\alpha-1} \exp\left\{tx - (\beta_k x)^\alpha\right\} \mathrm{d}x.$$

We can obtain $I_k(t)$ by using the Wright generalized hypergeometric function (Wright, 1935 [172]) defined by (1.12) in Chapter 1. The Wright function exists if $1 + \sum_{j=1}^{q} B_j - \sum_{j=1}^{p} A_j > 0$. For $\alpha > 1$, we have

$$
\begin{aligned}
I_k(t) &= \sum_{m=0}^{\infty} \frac{t^m}{m!} \int_0^\infty x^{m+\alpha-1} \exp\left\{-(\beta_k x)^\alpha\right\} \mathrm{d}x \\
&= \frac{1}{\alpha\beta_k^\alpha} \sum_{m=0}^{\infty} \frac{(t/\beta_k)^m}{m!}\,\Gamma\left(\frac{m}{\alpha} + 1\right) \\
&= \frac{1}{\alpha\beta_k^\alpha}\, {}_1\Psi_0\left[\begin{matrix} (1,\alpha^{-1}) \\ - \end{matrix}\; ; \frac{t}{\beta_k} \right].
\end{aligned} \tag{6.12}
$$

Combining (6.11) and (6.12), we obtain (provided that $\alpha > 1$)

$$M(t) = \alpha^{-1}\beta^{-\alpha} \sum_{k=0}^{\infty} \frac{v_k}{(k+1)^\alpha}\, {}_1\Psi_0\left[\begin{matrix} (1,\alpha^{-1}) \\ - \end{matrix}\; ; \frac{t}{\beta_k} \right]. \tag{6.13}$$

A second expression for $M(t)$ can be based on the Meijer G-function defined by (1.9) in Chapter 1. The Meijer G-function has several integrals with basic and special functions (Prudnikov *et al.*, 1986) [145].

For an arbitrary $g(\cdot)$ function, we use

$$\exp\{-g(x)\} = G_{0,1}^{1,0}\left(g(x)\,\middle|\, \begin{matrix} - \\ 0 \end{matrix}\right)$$

and then $I_k(t)$ becomes

$$I_k(t) = \int_0^\infty x^{\alpha-1}\,\exp(sx)\,G_{0,1}^{1,0}\left(\beta_k^\alpha\,x^\alpha\,\middle|\, \begin{matrix} - \\ 0 \end{matrix}\right)\mathrm{d}x.$$

Next, we assume $p \geq 1$ and $q \geq 1$ co-prime integers and $\alpha = p/q$. This requirement is not too restrictive, since every real number can be approached by a rational number.

By using equation (2.24.1.1) in Prudnikov *et al.* (1986, vol. 3) [145], we obtain

$$I_k(t) = \frac{p^{p/q-1/2}(-t)^{-p/q}}{(2\pi)^{(p+q)/2-1}}G_{q,p}^{p,q}\left(\frac{(\beta_k c)^q p^p}{(-t)^p q^q}\,\middle|\, \begin{matrix} \frac{1-p/q}{p}, \frac{2-p/q}{p}, \dots, \frac{p-p/q}{p} \\ 0, \frac{1}{q}, \dots, \frac{q-1}{q} \end{matrix}\right).$$

From (6.11) and the last equation, we have

$$M(t) = A(t)\sum_{k=0}^\infty v_k\,G_{q,p}^{p,q}\left(\frac{(\beta_k p/q)^q p^p}{(-t)^p q^q}\,\middle|\, \begin{matrix} \frac{1-p/q}{p}, \frac{2-p/q}{p}, \dots, \frac{p-p/q}{p} \\ 0, \frac{1}{q}, \dots, \frac{q-1}{q} \end{matrix}\right), \quad (6.14)$$

where $A(t) = p^{p/q-1/2}(-t)^{-p/q}/(2\pi)^{(p+q)/2-1}$. Equations (6.13) and (6.14) are the main results of this section. Note that the generating functions for the Kumaraswamy exponential, Weibull, exponentiated Weibull and exponentiated Rayleigh models can be obtained from the above expressions by choosing known values for the parameters.

6.1.4. Maximum Likelihood Estimation

We estimate the unknown parameters of the $\mathrm{KwW}(a, b, \alpha, \beta)$ model by maximum likelihood from complete samples. The log-likelihood function for the vector of parameters $\boldsymbol{\theta} = (a, b, \alpha, \beta)^\top$ from a sample $x_1, \dots, x_n$ of size n is

$$\ell(\boldsymbol{\theta}) = n\log(ab\alpha\beta^\alpha) + (\alpha-1)\sum_{i=1}^n\log(x_i) - \beta^\alpha\sum_{i=1}^n x_i^\alpha$$
$$+ (a-1)\sum_{i=1}^n\log(u_i) + (b-1)\sum_{i=1}^n\log(1-u_i^a),$$

where $u_i = 1 - \exp\{-(\beta x_i)^u\}$.

Further, the score components for the parameters α, β, a and b are

$$
\begin{aligned}
U_\alpha(\boldsymbol{\theta}) &= \frac{n}{\alpha} + r\log(\beta) + \sum_{i=1}^{n}\log(x_i) - \beta^\alpha \log(\beta)\sum_{i=1}^{n}x_i^\alpha - \beta^\alpha\sum_{i=1}^{n}x_i^\alpha \log(x_i) \\
&\quad + (a-1)\beta^\alpha\sum_{i\in F}\frac{x_i^\alpha \log(\beta x_i)(1-u_i)}{u_i} \\
&\quad -(b-1)a\beta^\alpha\sum_{i=1}^{n}\frac{x_i^\alpha \log(\beta x_i)u_i^{a-1}(1-u_i)}{(1-u_i^a)},
\end{aligned}
$$

$$
\begin{aligned}
U_\beta(\boldsymbol{\theta}) &= n\alpha\beta^{-1} - \alpha\beta^{\alpha-1}\sum_{i=1}^{n}x_i^\alpha + (a-1)\beta^{\alpha-1}\sum_{i=1}^{n}\frac{x_i^\alpha(1-u_i)}{u_i} - \\
&\quad (b-1)a\alpha\beta^{\alpha-1}\sum_{i\in F}\frac{x_i^\alpha u_i^{a-1}(1-u_i)}{(1-u_i^a)},
\end{aligned}
$$

$$
U_a(\boldsymbol{\theta}) = \frac{n}{a} + \sum_{i=1}^{n}\log(u_i) - (b-1)\sum_{i=1}^{n}\frac{u_i^a \log(u_i)}{(1-u_i^a)},
$$

$$
U_b(\boldsymbol{\theta}) = \frac{n}{b} + \sum_{i=1}^{n}\log(1-u_i^a).
$$

Then, the MLE $\widehat{\boldsymbol{\theta}}$ of $\boldsymbol{\theta}$ can be determined as the solution of the nonlinear equations $U_\alpha(\boldsymbol{\theta}) = 0$, $U_\beta(\boldsymbol{\theta}) = 0$, $U_a(\boldsymbol{\theta}) = 0$ and $U_b(\boldsymbol{\theta}) = 0$.

These equations can not be solved analytically. So, iterative methods such as Newton-Raphson algorithms, available in statistical platforms, are required to evaluate $\widehat{\boldsymbol{\theta}}$. For interval estimation and hypotheses tests on the model parameters, we can use the observed information matrix since the expected information matrix needs numerical integration methods.

The 4×4 total observed information matrix is $J(\boldsymbol{\theta}) = \{-J_{rs}\}$ (for $r, s \in \{a, b, \alpha, \beta\}$), whose second-order log-likelihood derivatives J_{rs} can be evaluated numerically.

Under general regularity conditions, the distribution of $(\widehat{\boldsymbol{\theta}} - \boldsymbol{\theta})$ can be approximated by the multivariate normal $N_4(\mathbf{0}, \boldsymbol{K}(\boldsymbol{\theta})^{-1})$, where $\boldsymbol{K}(\boldsymbol{\theta})$ is the total information matrix. So, the multivariate normal $N_4(\mathbf{0}, \boldsymbol{J}(\widehat{\boldsymbol{\theta}})^{-1})$ distribution can be adopted to construct confidence intervals for the model parameters.

The LR statistic can be used for testing if the fit of the KwW distribution is statistically superior to the fits of some of its special models. For example, we compare the KwW and EW models by testing $H_0 : b = 1$ versus $H_1 : b \neq 1$. The same can be done for the KwW and Weibull models by testing $H_0 : a = b = 1$ versus $H_1 :$ H_0 is false.

The LR statistic for testing $H_0 : b = 1$ *versus* $H : b \neq 1$ becomes

$$w = 2\{\ell(\widehat{a}, \widehat{b}, \widehat{\alpha}, \widehat{\beta}) - \ell(\widetilde{a}, 1, \widetilde{\alpha}, \widetilde{\beta})\},$$

where $\widehat{\alpha}$, $\widehat{\beta}$, $\widehat{a}$ and $\widehat{b}$ are the MLEs under H and $\widetilde{\alpha}$, $\widetilde{\beta}$ and $\widetilde{a}$ are the estimates under H_0.

6.1.5. Applications

In this section, we provide two examples using well-known data to prove empirically the advantages of the proposed model over other lifetime competing models.

Voltage Data

We compare the KwW, EW, EE and Weibull models to fit them to the times of failure and running times of devices from a field-tracking research. These data come from a larger system studied by Meeker and Escobar (1998, p. 383) [3]. Two supposed failure causes are considered for each unit: normal product wear and accumulation of random damages.

Table 6.1 gives the MLEs (and their corresponding SEs in parentheses) of the parameters and the values of the AIC, BIC and CAIC statistics for the competing distributions. We use the `NLMixed` computational procedure in `SAS` for the estimation process. The values in this table reveal that the KwW model has the lowest values for the AIC, BIC and CAIC statistics among all competing models, and then it could be chosen as the best model.

Test Stopped Data

We consider observations representing failure times, where the items are running under test stopped. The data are given by Murthy *et al.* (2004, p. 154) [57]. Table 6.2 lists the MLEs (and their SEs in parentheses) of the parameters and the values of the AIC, BIC and CAIC statistics for some fitted models. The algebraic calculations are performed in the `NLMixed` procedure in `SAS`.

In any case, these three statistics give the smaller values for the KwW model if compared with those of the EW, EE and Weibull distributions. Then, the KwW distribution is a very competitive model for lifetime data analysis.

Table 6.1: MLEs and respective SEs (given in parentheses) for the voltage data and the values of AIC, BIC and CAIC statistics.

Model	a	b	α	β	AIC	CAIC	BIC
KwW	0.0516	0.2288	7.7026	0.0043	352.3	353.9	357.9
	(0.0241)	(0.0905)	(0.2191)	(0.0003)			
EW	0.1334	1	7.2105	0.0031	359.6	360.5	363.8
	(0.0253)	-	(0.2009)	(0.0002)			
EE	1.1543	1	1	0.0062	374.2	374.7	377.0
	(0.2734)	-	-	(0.0014)			
Weibull	1	1	1.2650	0.0053	372.6	373.1	375.4
	-	-	(0.2044)	(0.0008)			

Table 6.2: MLEs of the model parameters for the test stopped data, their SEs (given in parentheses) and the AIC, BIC and CAIC statistics.

Model	a	b	α	β	AIC	CAIC	BIC
KwW	0.0663	0.1725	4.4200	0.1744	98.6	101.3	102.6
	(0.0044)	(0.1013)	(0.1411)	(0.0276)			
EW	0.2617	1	3.8910	0.1055	106.9	108.4	109.9
	(0.0628)	-	(0.1371)	(0.0118)			
EE	0.8377	1	1	0.1570	113.2	113.9	115.2
	(0.2300)	-	-	(0.0467)			
Weibull	1	1	1.0893	0.1719	113.5	114.2	115.5
	-	-	(0.2210)	(0.0361)			

6.2. KUMARASWAMY BURR XII

Zimmer *et al.* (1998) [173] introduced the three-parameter Burr XII (BXII) distribution, whose cdf and pdf (for $x > 0$) are

$$G(x; s, k, c) = 1 - \left[1 + \left(\frac{x}{s}\right)^c\right]^{-k} \tag{6.15}$$

and

$$g(x; s, k, c) = c\,k\,s^{-c}\,x^{c-1}\left[1 + \left(\frac{x}{s}\right)^c\right]^{-k-1}, \tag{6.16}$$

respectively, where $k > 0$ and $c > 0$ are shape parameters and $s > 0$ is a scale parameter. If $c > 1$, the density function is unimodal with mode at $x = s\left[(c-1)/(ck+1)\right]^{1/c}$ and is L-shaped when $c = 1$. If $ck > q$, the qth moment about zero is $\mu'_q = s^q\, k\, B(k - q\, c^{-1}, 1 + q\, c^{-1})$, where $B(\cdot, \cdot)$ is the beta function.

The BXII distribution is very adequate for modeling lifetime data and phenomenon with monotone failure rates. It includes as special models the logistic and Weibull distributions. The BXII distribution is not adequate for modeling data with non-monotone failure rates such as those having bathtub hazard and unimodal forms. Data with bathtub hazard forms present approximately flat middle parts and the corresponding densities have a positive anti-mode. For the unimodal failure forms, the hrf reaches a peak after some period and then decreases gradually.

The BXII cdf has a closed-form expression, which facilitates the computation of the percentiles and the likelihood function for censored data. Further, the BXII distribution has adequate tails for modeling failures that occur infrequently, very different from distributions based on exponential tails (Zimmer *et al.*, 1998 [173]).

The estimation of the parameters by maximum likelihood was addressed by Shao (2004) [174]. Further, Shao *et al.* (2004) [175] discussed other models for extremes, which use as basis the BXII model, and applications to flood frequency analysis. This model encompasses some properties of a large number of distributions (Soliman, 2005 [176]). Its versatility is quite appealing as a competing model for lifetime data. Furthermore, the log-Burr XII location-scale regression model (Silva *et al.* (2008) [177]) is an important alternative to the log-logistic regression model.

Next, we study a five-parameter *Kumaraswamy Burr XII* (KwBXII) distribution by inserting (6.15) in equation (6.1). The model was pioneered by Paranaíba *et al.* (2012) [117]. Its cumulative distribution (for $x > 0$) becomes

$$F(x) = 1 - \left[1 - \left\{1 - \left[1 + \left(\frac{x}{s}\right)^c\right]^{-k}\right\}^a\right]^b, \tag{6.17}$$

and its pdf can be expressed, from equations (6.2), (6.15) and (6.16), as

$$f(x) = a\,b\,c\,k\,s^{-c}x^{c-1}\left[1 + \left(\frac{x}{s}\right)^c\right]^{-k-1}\left\{1 - \left[1 + \left(\frac{x}{s}\right)^c\right]^{-k}\right\}^{a-1} \times$$

$$\left[1 - \left\{1 - \left[1 + \left(\frac{x}{s}\right)^c\right]^{-k}\right\}^a\right]^{b-1}, \quad x > 0. \tag{6.18}$$

Equation (6.18) includes as special models several well-known distributions. Clearly, the BXII model is the particular model when $a = b = 1$. For $b = 1$, it gives the exponentiated Burr XII (EBXII) distribution. We have a new model so-called the Kumaraswamy log-logistic (KwLL) distribution after the re-parametrization $s = m^{-1}$ and $k = 1$. The Kumaraswamy Pareto type II (KwPaII) and Pareto type II (PaII) distributions follow when $a = c = 1$ and $a = b = c = 1$, respectively. For $a = b = 1$, $s = m^{-1}$ and $k = 1$, the KwBXII distribution reduces to the log-logistic (LL) distribution. It converges to the KwW distribution when $k \to \infty$. Note that the KwBXII distribution is suitable for modeling comfortable unimodal-shaped hazard rates and for testing special models. The importance of the KwBXII distribution is noted by the extensive usage of the BXII distribution for modeling complex real data.

Let $X \sim \mathrm{KwBXII}(a, b, s, k, c)$ denote a random variable with density function (6.18). The KwBXII qf, say $x = Q(u)$, can be obtained by inverting (6.17) as

$$x = Q(u) = F^{-1}(u) = s \left\{ \left(1 - \left[1 - (1-u)^{1/b} \right]^{1/a} \right)^{-1/k} - 1 \right\}^{1/c}. \tag{6.19}$$

It is easy to generate X by taking u as a uniform random variable on the unit interval $(0, 1)$. Most of the results in this section follow Paranaíba *et al.* (2011) [117]. Paranaíba *et al.* (2013) [169] derived and studied other properties of the KwBXII model.

6.2.1. Expansion for the Density Function

First, if $|z| < 1$ and $b > 0$ is real non-integer, the power series holds

$$(1 - z)^{b-1} = \sum_{j=0}^{\infty} (-1)^j \binom{b-1}{j} z^j. \tag{6.20}$$

For $b > 0$ real non-integer, we can write

$$G(x)^{a-1} \{1 - G(x)^a\}^{b-1} = \sum_{i=0}^{\infty} w_i \, G(x)^{a(i+1)-1},$$

where $w_i = (-1)^i \binom{b-1}{i}$. If $\alpha > 0$ is real non-integer, an expanded form of $G(x)^\alpha$ can be obtained from (6.20) as

$$G(x)^\alpha = \sum_{r=0}^{\infty} s_r(\alpha) \, G(x)^r, \tag{6.21}$$

where

$$s_r(\alpha) = \sum_{m=r}^{\infty} (-1)^{r+m} \binom{\alpha}{m} \binom{m}{r}.$$

Further, we can write

$$G(x)^{a-1}\{1 - G(x)^a\}^{b-1} = \sum_{r=0}^{\infty} t_r\, G(x)^r, \tag{6.22}$$

where $t_r = t_r(a, b) = \sum_{i=0}^{\infty}(-1)^i \binom{b-1}{i} s_r(a(i+1)-1)$. Expansion (6.22) is used throughout this section.

From equations (6.2), (6.15), (6.16) and (6.22), the density function of X can be expressed as

$$f(x) = a\,b\,c\,k\,s^{-c}x^{c-1}\left[1 + \left(\frac{x}{s}\right)^c\right]^{-k-1} \sum_{r=0}^{\infty} t_r \left[1 - \left[1 + \left(\frac{x}{s}\right)^c\right]^{-k}\right]^r. \tag{6.23}$$

Using the binomial expansion and then interchanging the double sum $\sum_{r=0}^{\infty}\sum_{j=0}^{r}$ by $\sum_{j=0}^{\infty}\sum_{r=j}^{\infty}$, $f(x)$ can be written as

$$f(x) = \sum_{j=0}^{\infty} v_j\, g(x; s, k(j+1), c), \tag{6.24}$$

where the coefficients v_j are

$$v_j = v_j(a, b) = a\,b \sum_{r=j}^{\infty} \frac{(-1)^j}{(j+1)} \binom{r}{j} t_r,$$

and $g(x; s, k(j+1), c)$ denotes the BXII density with shape parameters $k(j+1)$ and c, and scale parameter s. Equation (6.24) reveals that the KwBXII density function is a linear combination of BXII densities. Clearly, $\sum_{j=0}^{\infty} v_j = 1$. Integrating (6.24) gives

$$F(x) = \sum_{j=0}^{\infty} v_j\, G(x; s, k(j+1), c). \tag{6.25}$$

Equations (6.24) and (6.25) are the main results of this section.

6.2.2.   Moments

For $n < c\,k$, the nth moment of X follows directly from (6.24) using a result in Zimmer *et al.* (1998) [173] as

$$\mu'_n = \mathrm{E}(X^n) = s^n\,k \sum_{j=0}^{\infty}(j+1)\,v_j\,B(k(j+1) - n\,c^{-1}, 1 + n\,c^{-1}). \tag{6.26}$$

The hth incomplete moment of X, $m_h(y)$, can be determined from (6.24) as

$$m_h(y) = c\,k\sum_{j=0}^{\infty}(j+1)\,v_j\int_0^y x^{h-1}\left(\frac{x}{s}\right)^c\left[1+\left(\frac{x}{s}\right)^{-k(j+1)-1}\right]\mathrm{d}x.$$

Setting $t=\left[1+\left(\frac{x}{s}\right)^c\right]^{-1}$, we can write

$$m_h(y) = k\,s^h\sum_{j=0}^{\infty}(j+1)\,v_j\int_0^{\left(\frac{s^c}{s^c+y^c}\right)}t^{k(j+1)-\frac{h}{c}-1}\,(1-t)^{\frac{h}{c}}dt$$

and then

$$m_h(y) = k\,s^h\sum_{j=0}^{\infty}(j+1)\,v_j\,B_{\frac{s^c}{s^c+y^c}}(k(j+1)-n\,c^{-1},1+n\,c^{-1}),\qquad(6.27)$$

where $B_z(a,b)$ is the incomplete beta function.

6.2.3. Generating Function

The mgf $M(t)$ of X can follow from (6.24) as

$$M(t) = \sum_{j=0}^{\infty}v_j\,M_{k(j+1)}(t),\qquad(6.28)$$

where $M_{k(j+1)}(t)$ is the mgf of the BXII$(s,k(j+1),c)$ distribution. For the case $t<0$, we obtain a simple expression for $M_k(t)$. We have

$$M_k(t) = c\,k\int_0^{\infty}\exp(yst)\,y^{c-1}\,(1+y^c)^{-(k+1)}dy.$$

We require the Meijer G-function defined in (1.9). This function encapsulates many integrals with basic and special functions (see Prudnikov *et al.*, 1986 [145]). For m and n positive integers, $\mu>-1$ and $p>0$, we can write

$$
\begin{aligned}
I\left(p,\mu,\frac{m}{n},\nu\right) &= \int_0^{\infty}\exp(-px)\,x^{\mu}\,(1+x^{\frac{m}{n}})^{\nu}\mathrm{d}x\\[2mm]
&= \frac{n^{-\nu}m^{\mu+\frac{1}{2}}}{(2\pi)^{\frac{(m-1)}{2}}\Gamma(-\nu)p^{\mu+1}}\times\\[2mm]
&\quad G_{n+m,n}^{n,n+m}\left(\frac{m^m}{p^m}\,\middle|\,\begin{matrix}\Delta(m,-\mu),\Delta(n,\nu+1)\\\Delta(n,0)\end{matrix}\right),\qquad(6.29)
\end{aligned}
$$

where

$$\Delta(n, a) = \frac{a}{n}, \frac{a+1}{n}, \cdots, \frac{a+n}{n}$$

(see Prudnikov *et al.*, 1992, page 21 [145]). We assume $c = m/n$, where m and n are positive integers. By using the integral (6.29), we have

$$M_n(t) = m\, I\left(-s\,t, \frac{m}{n} - 1, \frac{m}{n}, -n - 1\right).$$

Every positive real number can be approached by a rational number, so it is not a restrictive condition. Thus, the mgf of X follows from (6.24) (when $t < 0$) as

$$M(t) = m \sum_{j=0}^{\infty} v_j\, I\left(-s\,t, \frac{m}{n(j+1)} - 1, \frac{m}{n(j+1)}, -n(j+1) - 1\right). \qquad (6.30)$$

We can obtain simple expressions of $M_n(t)$ when $c = 1$ and $c = 2$, and, consequently, for $M(t)$, using equations (1) (on page 16) and (2) (on page 20) of the book by Prudnikov *et al.* (1992) [145].

6.2.4. Estimation

Maximum Likelihood Estimation

Let $T_i \sim \mathrm{KwBXII}(a, b, s, k, c)$ and $\boldsymbol{\theta} = (a, b, s, k, c)^{\top}$ be the parameter vector. Censored data are common in reliability and survival analysis. A very practical method considers that each individual i has a lifetime T_i and a censoring time C_i, where both random variables are independent. The distribution of C_i is assumed independent of the unknown parameters of T_i. Further, we consider n independent observations $X_i = \min(T_i, C_i)$ (for $i = 1, \ldots, n$).

Parametric approach for such data is usually based on likelihood and asymptotic theories. The censored log-likelihood $l(\boldsymbol{\theta})$ for the model parameters is

$$
\begin{aligned}
\ell(\boldsymbol{\theta}) =\ & r\left[\log(a) + \log(b) + \log(c) + \log(k) - c\log(s)\right] \\
& + (c-1)\sum_{i \in F}\log(x_i) + (a-1)\sum_{i \in F}\log(u_i) + b\sum_{i \in C}\log(1 - u_i^a) \\
& + \left(\frac{k+1}{k}\right)\sum_{i \in F}\log(1 - u_i) + (b-1)\sum_{i \in F}\log(1 - u_i^a), \qquad (6.31)
\end{aligned}
$$

where r is the failures number, F and C denote the uncensored and censored sets of observations, respectively, and $u_i = 1 - \left[1 + \left(\frac{x_i}{s}\right)^c\right]^{-k}$. The components of the score function are

$$U_a(\boldsymbol{\theta}) = \frac{r}{a} + \sum_{i \in F} \log(u_i) - (b-1) \sum_{i \in F} \frac{(u_i)^a \log(u_i)}{(1-u_i^a)} - b \sum_{i \in C} (u_i)^a \log(u_i),$$

$$U_b(\boldsymbol{\theta}) = \frac{r}{b} + \sum_{i \in F} \log(1 - u_i^a) + \sum_{i \in C} (1 - u_i^a),$$

$$U_s(\boldsymbol{\theta}) = \frac{ck}{s} \sum_{i \in F} \left(\frac{x_i}{s}\right)^c (1 - u_i)^{\frac{1}{k}} \left[1 + (a-1)\frac{(1-u_i)}{(u_i)} - a(b-1)\frac{(u_i)^{a-1}(1-u_i)}{(1-u_i^a)}\right]$$
$$+ \frac{(c-1) - rc}{s} - \frac{abkc}{s} \sum_{i \in C} (u_i)^{a-1}(1-u_i)^{\frac{k+1}{k}} \left(\frac{x_i}{s}\right)^c,$$

$$U_k(\boldsymbol{\theta}) = \frac{r}{k} + \frac{1}{k} \sum_{i \in F} \log(1 - u_i) \left[1 - (a-1)\frac{(1-u_i)}{(u_i)} + a(b-1)\frac{(u_i)^{a-1}(1-u_i)}{(1-u_i^a)}\right]$$
$$+ \frac{ab}{k} \sum_{i \in C} (u_i)^{a-1}(1-u_i) \log(1-u_i)$$

and

$$U_c(\boldsymbol{\theta}) = \frac{r}{c} - r \log(s) + \sum_{i \in F} \log(x_i) + \sum_{i \in F} (1 - u_i)^{\frac{1}{k}} \left(\frac{x_i}{s}\right)^c \log\left(\frac{x_i}{s}\right)$$
$$\times \left[-(k+1) + k(a-1)\frac{(1-u_i)}{(u_i)} - ak(b-1)\frac{(u_i)^{a-1}(1-u_i)}{(1-u_i^a)}\right]$$
$$- abk \sum_{i \in C} (u_i)^{a-1}(1-u_i)^{\frac{k+1}{k}} \left(\frac{x_i}{s}\right)^c \log\left(\frac{x_i}{s}\right).$$

The MLE $\widehat{\boldsymbol{\theta}}$ of $\boldsymbol{\theta}$ can be obtained by solving simultaneously the nonlinear equations $U_a(\boldsymbol{\theta}) = 0$, $U_b(\boldsymbol{\theta}) = 0$, $U_s(\boldsymbol{\theta}) = 0$, $U_k(\boldsymbol{\theta}) = 0$ and $U_c(\boldsymbol{\theta}) = 0$.
There is no closed-form expressions for the MLEs and therefore nonlinear optimization algorithms should be used such as the Newton-Raphson iterative method to obtain $\widehat{\boldsymbol{\theta}}$. We adopt the `MaxBFGS` sub-routine in the programming matrix language `Ox` (Doornik, 2007), although there are others routines a-vailable for numerical maximization of (6.31) given in the R program (`optim` function) and SAS (`PROC NLMIXED`).
For making inference on the parameters, we can adopt the observed infor-mation matrix because the expected information matrix depends on integra-tion methods. The 5×5 observed information matrix is $J(\boldsymbol{\theta}) = \{-J_{rs}\}$ (for $r, s \in \{a, b, s, c, k\}$), whose second-order log-likelihood derivatives J_{rs} can be assessed numerically.
Under standard regularity conditions, the asymptotic distribution of $(\widehat{\boldsymbol{\theta}} - \boldsymbol{\theta})$ is $N_5(\mathbf{0}, \boldsymbol{K}(\boldsymbol{\theta})^{-1})$, where $\boldsymbol{K}(\boldsymbol{\theta})$ is the expected information matrix. We can

construct approximate confidence intervals for the model parameters based on $N_5(\mathbf{0}, J(\widehat{\boldsymbol{\theta}})^{-1})$ distribution, where $J(\widehat{\boldsymbol{\theta}})$ is the observed information matrix at $\widehat{\boldsymbol{\theta}}$. The LR statistics can be used for testing if the KwBXII distribution is statistically superior to some of its special models for the data under study.

A Bayesian Analysis

Bayesian analysis is an alternative procedure for estimating the parameters of a statistical model. Bayesian updating is particularly important in the dynamic analysis of a sequence of data. It embodies previous knowledge about the parameters based on informative priori density functions. The prior information can be obtained through a posterior marginal distribution. In this method there are two important aspects: the first one refers to the choice of the marginal posterior distribution, and the second one is to determine the moments of interest. These problems can be solved by Markov Chain Monte Carlo (MCMC) simulation approaches, which include the Metropolis-Hastings algorithm and the Gibbs sampler.

We assume that the unknown parameters are independent having joint prior density function

$$\pi(a, b, s, k, c) \propto \pi(a) \times \pi(b) \times \pi(s) \times \pi(k) \times \pi(c), \tag{6.32}$$

where $a \sim \Gamma(\alpha_1, \beta_1)$, $b \sim \Gamma(\alpha_2, \beta_2)$, $s \sim \Gamma(\alpha_3, \beta_3)$, $k \sim \Gamma(\alpha_4, \beta_4)$ and $c \sim \Gamma(\alpha_5, \beta_5)$. Here, $\Gamma(\alpha_i, \beta_i)$ indicates a gamma distribution with mean α_i/β_i, variance α_i/β_i^2 and density function

$$f(\upsilon; \alpha_i, \beta_i) = \frac{\beta_i^{\alpha_i} \upsilon^{\alpha_i-1} \exp(-\beta_i \upsilon)}{\Gamma(\alpha_i)},$$

where $\upsilon > 0$, $\alpha_i > 0$ and $\beta_i > 0$. All hyper-parameters are well-defined. Combining (6.31) and (6.32), the joint posterior distribution for a, b, s, k and c reduces to

$$\pi(a, b, s, k, c|y) \propto \left(a\,b\,c\,k\,s^{-c}\right)^r \prod_{i \in F} x_i^{c-1} u_i^{a-1} (1 + u_i)^{\frac{k+1}{k}} \left(1 - u_i^a\right)^{b-1}$$

$$\times \prod_{i \in C} \left(1 - u_i^a\right)^b \pi(a, b, s, k, c). \tag{6.33}$$

The integration of the the joint posterior density (6.33) can not be tractable analytically and the inference is based on MCMC simulation methods. The full conditional distributions of the unknown parameters are

$$\pi(a|y, b, s, k, c) \propto a^r \prod_{i \in F} u_i^a \left(1 - u_i^a\right)^{b-1} \prod_{i \in C} \left(1 - u_i^a\right)^b \times \pi(a),$$

$$\pi(b|y, a, s, k, c) \propto b^r \prod_{i \in F}(1 \quad u_i^a)^b \prod_{i \in C}(1 - u_i^a)^b \times \pi(b),$$

$$\pi(s|y, a, b, k, c) \propto s^{-cr} \prod_{i \in F} u_i^{a-1}(1 + u_i)^{\frac{k+1}{k}}(1 - u_i^a)^{b-1} \prod_{i \in C}(1 - u_i^a)^b \times \pi(s),$$

$$\pi(k|y, a, b, s, c) \propto k^r \prod_{i \in F} u_i^{a-1}(1 + u_i)(1 - u_i^a)^{b-1} \prod_{i \in C}(1 - u_i^a)^b \times \pi(k),$$

and

$$\pi(c|y, a, b, s, k) \propto \frac{1}{s^{cr}} \prod_{i \in F} x_i^c u_i^{a-1}(1 + u_i)^{\frac{k+1}{k}}(1 - u_i^a)^{b-1} \prod_{i \in C}(1 - u_i^a)^b \times \pi(c).$$

The full conditional distributions for the model parameters a, b, s, k and c do not have closed-forms and then we use the Metropolis-Hastings algorithm. The MCMC calculations are executed in the R software.

6.2.5. Simulation Studies

We carry out a simulation study for sample sizes $n = 50, 100, 200$ and 300 to verify the adequacy of the MLEs of the KwBXII parameters. We simulate 2,000 samples by taking the true parameter values: $a = 6.0$, $b = 11.0$, $s = 0.1$, $k = 0.6$ and $c = 50$. The simulation results listed in Table 6.3 reveal that the root mean squared errors (RMSEs) of the estimates decay toward zero when the sample size increases. The mean estimates tend to the true parameter values when n increases, as expected. Both facts are in agreement with first-order asymptotic theory.

Some simulations are performed for $n = 200$ and 300 to compare the Burr XII and KwBXII distributions. We simulate 2,000 samples for the true parameter values $a = 6.0$, $b = 11.0$, $s = 0.1$, $k = 0.6$ and $c = 1.2$. For most cases, the simulations given in Table 6.4 indicate that the RMSEs of the estimates decay to zero when n increases. The RMSEs for the estimates of the KwBXII parameters are small if compared with those of the BXII distribution. Further, the mean estimates tend to the true parameter values when n increases.

6.2.6. Applications

In this section, we compare the KwBXII distribution with three of its special distributions (the LL, KwW and BXII models) by means of three real data sets. The first and second data sets are not censored but the third one presents

Table 6.3: Monte Carlo simulation results: mean estimates and RMSEs of $\hat{a}$, $\hat{b}$, $\hat{s}$, $\hat{k}$ and $\hat{c}$.

n	Parameter	Mean	RMSE
50	a	6.0007	3.6127
	b	10.8610	2.2146
	s	0.1004	2.2933×10^{-6}
	k	0.6777	0.0380
	c	50.2101	1.1468
100	a	5.6724	1.6672
	b	10.9042	0.9115
	s	0.1004	1.3264×10^{-6}
	k	0.6437	0.0128
	c	50.0635	0.4145
200	a	5.6723	0.5913
	b	10.9274	0.4685
	s	0.1003	5.4350×10^{-7}
	k	0.6278	0.0049
	c	50.0165	0.2712
300	a	5.5824	0.6451
	b	10.9699	0.1308
	s	0.1004	6.4989×10^{-7}
	k	0.6274	0.0040
	c	49.9717	0.2202

censored data. For each case, we estimate the parameters by maximum likelihood (see Section 6.2.4) using the NLMixed subroutine in SAS. Next, we report the data sets and provide the MLEs (and the corresponding SEs in parentheses) of the model parameters and the values of the AIC, CAIC and BIC statistics. The lower values of these statistics correspond to better fits.

(i) Fibre data
The first data set given by Smith and Naylor (1987) [178] and Cordeiro and Lemonte (2011) [118] constitutes of $n = 51$ strengths of 1.5 cm glass fibres measured at the National Physical Laboratory, England.

(ii) Stress data
The second data set given by Andrews and Herzberg (1985) [179] and Cooray and Ananda (2008) [180] constitutes of $n = 101$ observations on the stress-

Table 6.4: Monte Carlo simulation results: mean estimates and RMSEs of $\hat{a}$, $\hat{b}$, $\hat{s}$, $\hat{k}$ and $\hat{c}$.

n	Parameter	BXII		KBXII	
		Mean	RMSE	Mean	RMSE
	a			6.4650	4.8374
	b			11.6956	7.9843
200	s	0.3667	0.0753	0.1820	0.0758
	k	1.2586	0.6248	0.6650	0.3933
	c	3.0847	3.6686	1.7610	1.6368
	a			6.2947	2.2593
	b			11.4752	8.2817
300	s	0.3622	0.0713	0.1542	0.0292
	k	1.2240	0.5004	0.6385	0.2275
	c	3.0716	3.5782	1.5897	0.7976

rupture life of kevlar 49/epoxy strands, which are exposed to constant maintained pressure at the 90% stress level until all had failed. The aim is to obtain complete data with precise failure times.

(iii) AIDS data

The third data set refers to the times to serological reversal of $n = 143$ children who had HIV by vertical transmission, where the mothers were not treated (Silva, 2004 [182]; Perdoná, 2006 [183]). It is possible serological reversal in children of HIV-contaminated mothers when HIV antibodies vanish from the blood in an individual who tested positive for HIV infection. We have 16.8% and 82.2% of censored and uncensored observations, respectively. We evaluate the descriptive statistics only for the uncensored data.

Table 6.5 gives a descriptive summary for each data set. The fibre sample has negative skewness and the lowest variability in the data. The stress observations present positive skewness and kurtosis. The AIDS data have more variability in uncensored observations. Further, we evaluate the MLEs and the

Table 6.5: Descriptive statistics.

Data	Mean	Median	Mode	SD	Skewness	Kurtosis	Min.	Max.
Fibres	1.44	1.52	1.61	0.33	-0.66	1.02	0.55	2.24
Stress	1.03	0.80	0.02^a	1.12	3.05	14.51	0.01	7.89
AIDS	460.33	461	334	188.9	0.17	0.64	5	1021

^{a} There are several modes

AIC, BIC and CAIC statistics for some competing models. The estimates of s, k and c from the fitted BXII distribution are taking as the initial values for the numerical iterative method. We fit the KwBXII distribution to the three data sets and compare the results with those obtained from other competing models.

The BW distribution (Famoye *et al.*, 2005) is an important model for lifetime data. Its pdf (for $x > 0$) with four parameters $a > 0$, $b > 0$, $\alpha > 0$ and $\gamma > 0$ is

$$f(x) = \frac{\gamma}{\alpha^\gamma B(a, b)} x^{\gamma-1} \exp\left[-b\left(\frac{x}{\alpha}\right)^\gamma\right] \left\{1 - \exp\left[-\left(\frac{x}{\alpha}\right)^\gamma\right]\right\}^{a-1}.$$

An alternative model for these data is the KwW distribution (Cordeiro *et al.*, 2010 [84]), whose density function is

$$f(x) = a\,b\,c\,\lambda^c\,x^{c-1} \exp\{-(\lambda x)^c\} \left[1 - \exp\{-(\lambda x)^c\}\right]^{a-1}$$
$$\times \left\{1 - \left[1 - \exp\{-(\lambda x)^c\}\right]^a\right\}^{b-1},$$

where $a > 0$ and $b > 0$ are two extra shape parameters for the Weibull model. For $a = b = 1$, the Weibull distribution follows as a special case.

The lower values of the AIC, BIC and CAIC statistics are not necessarily a consequence of the extra parameter of the KwBXII model. The results are given in Table 6.6. Note that the three information criteria agree on the ranking of the models in each case. The lowest values for the three information criteria are provided by the fitted KwBXII distribution for the three data sets. Further, the BIC statistic indicates that the KwBXII and BXII distributions are the best models for the stress and AIDS data.

We can perform formal tests on the third skewness parameter in the KwBXII model based on the LR statistics. The results of the tests for the three data sets are listed in Table 6.7. For the fibre, stress and AIDS data, we reject the null hypotheses for the six LR tests in favor of the KwBXII distribution. This fact makes evident the potential of the three skewness parameters for modeling

Table 6.6: MLEs of the model parameters for the data, the corresponding SEs (given in parentheses) and the AIC, BIC and CAIC statistics.

Fibres	a	b	s	k	c	AIC	CAIC	BIC
KwBXII	0.1322	0.8944	1.6440	0.3358	32.3362	25.2	26.5	34.9
	(0.0333)	(0.4094)	(0.0474)	(0.2073)	(1.4639)			
BXII	1	1	2.3424	10.4426	5.6369	33.2	33.7	39.0
	-	-	(0.9581)	(1.6855)	(0.9406)			
	m	c						
LL	0.6865	7.5388				40.8	41.1	44.7
	(0.0215)	(0.9255)						
	a	b	α	γ				
BW	0.8470	5.9117	2.2035	5.7680		35.4	36.2	43.1
	(0.4142)	(5.5632)	(1.6975)	(1.7535)				
	a	b	λ	c				
KwW	0.0265	0.1089	0.8514	5.4060		70.3	71.2	78.0
	(0.0113)	(0.0169)	(0.0321)	(0.3164)				
Stress	a	b	s	k	c	AIC	CAIC	BIC
KwBXII	0.0986	0.8854	1.6847	0.3943	6.8188	207.5	208.1	220.6
	(0.0178)	(0.2856)	(0.2641)	(0.1436)	(0.2888)			
BXII	1	1	2217.90	1313.27	0.9311	212.3	212.5	220.1
	-	-	(296.98)	(542.25)	(0.0727)			
	m	c						
LL	1.5926	1.2810				229.4	229.5	234.7
	(0.2148)	(0.1079)						
	a	b	α	γ				
BW	0.7766	2.4128	2.9155	1.0864		213.9	214.3	224.4
	(0.3342)	(1.3841)	(1.3152)	(0.3050)				
	a	b	λ	c				
KwW	0.7303	0.2354	3.6684	1.0300		213.6	214.0	224.1
	(0.2399)	(0.0448)	(1.6456)	(0.1360)				
AIDS	a	b	s	k	c	AIC	CAIC	BIC
KwBXII	0.4834	0.1549	1220.12	64.9806	2.9577	1628.6	1629.0	1643.4
	(0.1100)	(0.0161)	(42.9089)	(4.1754)	(0.1443)			
BXII	1	1	14480	2408.38	2.3642	1632.5	1632.7	1641.4
	-	-	(444.84)	(1143.61)	(0.1704)			
	m	c						
LL	0.002198	3.2885				1650.4	1650.5	1656.3
	(0.000099)	(0.2586)						
	a	b	α	γ				
BW	0.7159	0.7455	537.00	2.8720		1633.1	1633.4	1645.0
	(0.2276)	(1.1088)	(88.81)	(0.7676)				
	a	b	λ	c				
KwW	0.7367	15.0306	0.00057	3.1385		1634.4	1634.7	1646.2
	(0.1074)	(0.0048)	(0.000068)	(2.8669)				

Table 6.7: LR tests.

Fibres	Hypotheses	w	p-value
KwBXII vs BXII	$H_0 : a = b = 1$ vs $H_1 : H_0$ is false	12.0	0.00248
KwBXII vs LL	$H_0 : a = b = k = 1$ vs $H_1 : H_0$ is false	21.6	<0.0001
Stress	Hypotheses	w	p-value
KwBXII vs BXII	$H_0 : a = b = 1$ vs $H_1 : H_0$ is false	8.8	0.01227
KwBXII vs LL	$H_0 : a = b = k = 1$ vs $H_1 : H_0$ is false	27.9	<0.0001
AIDS	Hypotheses	w	p-value
KwBXII vs BXII	$H_0 : a = b = 1$ vs $H_1 : H_0$ is false	7.9	0.01925
KwBXII vs LL	$H_0 : a = b = k = 1$ vs $H_1 : H_0$ is false	27.8	<0.0001

real survival data.

Bayesian analysis. For these data sets, the independent priors to run the Metropolis-Hastings algorithm are: $a \sim \Gamma(0.01, 0.01)$, $b \sim \Gamma(0.01, 0.01)$, $s \sim \Gamma(0.01, 0.01)$, $k \sim \Gamma(0.01, 0.01)$ and $c \sim \Gamma(0.01, 0.01)$. For these prior densities, we provide two parallel and independent sequences of the Metropolis-Hastings with size 50,000 for each parameter. We delete the first 5,000 iterations in order to eliminate the effects of the initial values. Further, we consider a spacing of size 10, obtaining a sample of size 4,500 from each chain in order to avoid correlation problems.

The convergence of the Metropolis-Hastings algorithm follows from the procedures introduced by Cowles and Carlin (1996) [184]. Then, we consider the between and within sequence information based on the method pioneered by Gelman and Rubin (1992) [185] to achieve the potential scale reduction, say $\widehat{R}$. For each case, these values approach one, thus demonstrating the convergence of the chain. The posterior summaries for the parameters of the KwBXII distribution for the current data sets are reported in Table 6.8. The values for the posteriori means (Table 6.8) as expected are close to the MLEs determined from the KwBXII distribution reported in Table 6.6. Note that the acronym SD refers to the standard deviation from the posterior distributions of the parameters and HPD means the 95% highest posterior density intervals.

Table 6.8: Posterior summaries for the parameters of the KwBXII model for the data sets.

Fibres				
Parameter	Mean	SD	HPD (95%)	$\widehat{R}$
a	0.1324	0.0410	(0.0542, 0.2130)	1.0011
b	0.8938	0.0302	(0.8359, 0.9533)	1.0021
s	1.6439	0.0102	(1.6237, 1.6623)	1.0019
k	0.3359	0.0221	(0.2938, 0.3793)	1.0017
c	32.3364	0.0908	(32.1581, 32.5112)	1.0023

Stress				
Parameter	Mean	SD	HPD (95%)	$\widehat{R}$
a	0.0990	0.0255	(0.0510, 0.1495)	1.0014
b	0.8850	0.0171	(0.8523, 0.9188)	1.0020
s	1.6845	0.0301	(1.6238, 1.7398)	1.0019
k	0.3945	0.0403	(0.3180, 0.4735)	1.0017
c	6.8195	0.0504	(6.7196, 6.9141)	1.0025

AIDS				
Parameter	Mean	SD	HPD (95%)	$\widehat{R}$
a	0.4852	0.0914	(0.3137, 0.6665)	0.9996
b	0.1535	0.0513	(0.0516, 0.2520)	1.0001
s	1220.28	0.6969	(1218.91, 1221.62)	1.0011
k	64.9848	0.5050	(64.0134, 65.9707)	1.0009
c	2.9591	0.1004	(2.7654, 3.1586)	0.9999

6.3. KUMARASWAMY GUMBEL

The Gumbel distribution has been extensively applied in many research fields such as engineering, space and software reliability, queues in supermarkets, among others. A recent book by Kotz and Nadarajah [185] lists over fifty applications of this distribution. We investigate some mathematical properties of the *Kumaraswamy Gumbel* (KwGum) distribution in order that it will attract more complex applications in lifetime research and reliability engineering. This model was investigated by Cordeiro *et al.* (2012) [170].

Henceforth, we express the cdf and pdf of the Gumbel model by $G_{\mu,\sigma}(x) = e^{-u}$ and $g_{\mu,\sigma}(x) = \sigma^{-1} u e^{-u}$, respectively, where $x, \mu \in \mathbb{R}$, $\sigma > 0$ and $u = \exp\{-(x - \mu)/\sigma\}$. The cdf and pdf of the KwGum distribution are

$$F(x) = 1 - \{1 - \exp(-au)\}^{b} \qquad (6.34)$$

and

$$f(x) = a\, b\, \sigma^{-1}\, u\, \exp(-au)\, \{1 - \exp(-au)\}^{b-1}, \qquad (6.35)$$

respectively. The KwGum density function (6.35) is more flexible than the Gumbel distribution and simpler than the pdf of the beta Gumbel distribution (Nadarajah and Kotz, 2004 [158]). This extension can allow for greater flexibility of its tails.

The Gumbel model provides good fits to time series of the squares of the extreme wind speeds. Then, improved estimation and prediction can be provided by capturing the tail behavior more accurately. The proposed model (6.35) is a good alternative to achieve this.

Hereafter, a random variable with density (6.35) is denoted by $X \sim$ Kw-Gum(a, b, μ, σ). The KwGum qf is obtained by inverting (6.34)

$$x = Q(u) = F^{-1}(u) = \mu - \sigma \log\left\{\log\left[1 - (1 - u)^{1/b}\right]^{1/a}\right\}. \qquad (6.36)$$

Then, we can generate KwGum random numbers from (6.36) by $X = Q(U)$, where U is uniformly distributed on the interval $(0, 1)$.

6.3.1. Distribution and Density Functions

First, we provide useful expansions for the cumulative and density functions of the KwGum distribution in order to obtain some of its mathematical properties. Consider the power series (for α real non-integer and $|z| < 1$)

$$(1 + z)^{\alpha} = \sum_{j=0}^{\infty} \frac{\Gamma(\alpha + 1)}{\Gamma(\alpha - j + 1)} \frac{z^{j}}{j!}.$$

For b real non-integer, the KwGum cdf (6.34) can be reduced to

$$F(x) = 1 - \sum_{k=0}^{\infty} \frac{(-1)^k \; \Gamma(b+1) \; \exp(-kau)}{\Gamma(b-k+1) \; k!}, \tag{6.37}$$

whereas the KwGum pdf (6.35) can be expressed as

$$f(x) = a \; \sigma^{-1} \; \Gamma(b+1) \sum_{k=0}^{\infty} \frac{(-1)^k \; u \; \exp\{-(k+1) \; a \; u\}}{\Gamma(b-k) \; k!}. \tag{6.38}$$

These equations are the most important results of this section. They can be written as simple linear combinations of Gumbel cdfs and pdfs.

Further, the asymptotes of $f(x)$ and $F(x)$ when $x \to 0, \infty$ are given by

$$f(x) \sim \sigma^{-1} abu \exp(-au) \quad \text{as} \quad u \to \infty,$$

$$f(x) \sim \sigma^{-1} a^b b u^b \quad \text{as} \quad u \to 0,$$

$$F(x) \sim b \exp(-au) \quad \text{as} \quad u \to \infty,$$

$$1 - F(x) \sim (au)^b \quad \text{as} \quad u \to 0.$$

Note that the upper tail of $f(x)$ is of exponential type while the lower tail is of double exponential type.

6.3.2. Shapes

The first derivative of $\log[f(x)]$ for the KwGum density becomes

$$\frac{d \log f(x)}{dx} = -\frac{u}{\sigma} \left\{ \frac{1}{u} - a + \frac{a(b-1)}{\exp(au) - 1} \right\},$$

where $u = \exp\{-(x - \mu)/\sigma\}$. So, the modes of $f(x)$ are the roots of the equation

$$\frac{a(b-1)}{\exp(au) - 1} = a - \frac{1}{u}. \tag{6.39}$$

Note that there may be more than one root for (6.39). If $x = x_0$ is a root of (6.39), then it coincides to a local maximum, a local minimum or a point of inflexion depending on whether $\lambda(x_0) < 0$, $\lambda(x_0) > 0$ or $\lambda(x_0) = 0$, where $\lambda(x) = d^2 \log f(x)/dx^2$ is

$$\lambda(x) = -\frac{u^2}{\sigma^2} \left\{ \frac{1}{u^2} + \frac{a^2(b-1)\exp(au)}{[\exp(au) - 1]^2} \right\} + \frac{u}{\sigma^2} \left\{ \frac{1}{u} - a + \frac{a(b-1)}{\exp(au) - 1} \right\}.$$

6.3.3.   Moments

Let X have pdf (6.35). Setting $x = \mu - \sigma \log(u)$, the nth moment of X can be determined from (6.38) as

$$E(X^n) = a\Gamma(b+1) \sum_{k=0}^{\infty} \frac{(-1)^k}{\Gamma(b-k)\,k!} \int_0^\infty \{\mu - \sigma \log(u)\}^n \, \exp\{-(k+1)au\}\,du.$$

Based on the binomial theorem, we can write $E(X^n)$ as

$$E(X^n) = a\Gamma(b+1) \sum_{k=0}^{\infty} \sum_{j=0}^{n} \frac{(-1)^{k+j}\,\sigma^j\,\mu^{n-j}}{\Gamma(b-k)\,k!}$$
$$\times \binom{n}{j} \int_0^\infty \log^j(u) \exp\{-(k+1)au\}du.$$

For $a > 0$ and j a non-negative integer, let

$$I(j,a) = \int_0^\infty \log^j(u)\,e^{-au}du,$$

which can be determined by equation (2.6.21.1) in Prudnikov *et al.* (1986) [145] as

$$I(j,a) = \left(\frac{\partial}{\partial \alpha}\right)^j \left[a^{-\alpha}\Gamma(\alpha)\right]\Big|_{\alpha=1}.$$

Then, the nth moment of X becomes

$$E(X^n) = a\Gamma(b+1) \sum_{k=0}^{\infty} \sum_{j=0}^{n} \frac{(-1)^{k+j}\,\sigma^j\,\mu^{n-j}}{\Gamma(b-k)\,k!} \binom{n}{j} I(j,(k+1)u).$$

For $n = 1$, we obtain the special cases to evaluate the mean

$$I(0,a) = \frac{1}{a} \quad \text{and} \quad I(1,a) = \frac{\psi(1) - \log(a)}{a},$$

where $\psi(\cdot)$ is the digamma function. Finally,

$$E(X) = \Gamma(b+1) \sum_{k=0}^{\infty} \frac{(-1)^k\,[\mu - \sigma\{\psi(1) - \log[(k+1)a]\}]}{\Gamma(b-k)\,(k+1)!}.$$

The skewness and kurtosis of X are mostly controlled by a and b.

6.3.4.   Generating Function

Let X be a random variable having the KwGum density function (6.35). We provide two formal expressions for the mgf of X, $M(t) = E[\exp(tX)]$. By changing $x = \mu - \sigma \log(u)$ to u in equation (6.38), we obtain

$$
\begin{aligned}
M(t) &= \sigma^{-1}a\Gamma(b+1)\sum_{k=0}^{\infty}\frac{(-1)^k}{\Gamma(b-k)\ k!}\int_{-\infty}^{\infty}\exp(tx)u\exp\left\{-(k+1)au\right\}\mathrm{d}x \\
&= ae^{t\mu}\ \Gamma(b+1)\sum_{k=0}^{\infty}\frac{(-1)^k}{\Gamma(b-k)\ k!}\int_{0}^{\infty}u^{-\sigma t}\ \exp\left\{-(k+1)au\right\}\mathrm{d}u.
\end{aligned}
$$

By using equation (2.3.3.1) in Prudnikov *et al.* (1986) [145], $M(t)$ is

$$
M(t) = a^{\sigma t}\ e^{t\mu}\ \Gamma(b+1)\ \Gamma(1-\sigma t)\sum_{k=0}^{\infty}\frac{(-1)^k\ (k+1)^{\sigma t-1}}{\Gamma(b-k)\ k!}. \tag{6.40}
$$

We now provide an alternative expression for the mgf of X. For $b > 0$ real non-integer, we have

$$
G(x)^{a-1}\ \{1-G(x)^a\}^{b-1} = \sum_{i=0}^{\infty}q_i\ G(x)^{a(i+1)-1}, \tag{6.41}
$$

where the coefficients are $q_i = (-1)^i\binom{b-1}{i}$.
Then, using (6.41), $M(t)$ can be expressed as

$$
M(t) = a\ b\sum_{i=0}^{\infty}q_i\ \rho\left(t, a(i+1)-1\right), \tag{6.42}
$$

where $\rho(t,a) = \int_{-\infty}^{\infty}e^{tx}G(x)^a g(x)\mathrm{d}x$ can be determined from the Gumbel qf $Q(u) = G^{-1}(u) = \mu - \sigma\log[-\log(u)]$ as

$$
\rho(t,a) = \int_{0}^{1}\exp[t\,Q(u)]\,u^a\mathrm{d}u. \tag{6.43}
$$

For a positive integer a, the interpretation of the above integral is very easy. Except for a missing multiplier $a+1$, equation (6.43) is the mgf of the maximum in a random sample of size $a + 1$ from the G distribution. So, we can write from (6.43)

$$
\rho(t,a) = e^{t\mu}\int_{0}^{1}\log^{-\sigma\,t}\left(\frac{1}{u}\right)u^a\mathrm{d}u.
$$

Using equation (2.6.3.1) in Prudnikov *et al.* (1986) [145], we obtain (for $t < \sigma^{-1}$)

$$\rho(t,a) = a^{\sigma t - 1} \, e^{t\mu} \, \Gamma(1 - \sigma t).$$

Then, equation (6.42) becomes

$$M(t) = a \, b \, e^{t\mu} \, \Gamma(1 - \sigma t) \sum_{i=0}^{\infty} q_i \left[(i+1)a - 1 \right]^{\sigma t - 1}, \qquad (6.44)$$

which holds for $\sigma \, t < 1$.
Equations (6.40) and (6.44) are the main results of this section.

6.3.5.   Maximum Likelihood Estimation

We estimate the parameters of the KwGum distribution by maximum likelihood. Let $x_1, \ldots, x_n$ be a random sample from X and $\boldsymbol{\theta} = (a, b, \mu, \sigma)^{\top}$ the vector of the model parameters. The log-likelihood function for $\boldsymbol{\theta}$ is

$$
\begin{aligned}
\ell(\boldsymbol{\theta}) \;=\; & n \log(ab) - n \log(\sigma) - \sum_{i=1}^{n} \left(\frac{x_i - \mu}{\sigma} \right) - a \sum_{i=1}^{n} \exp\left[-\left(\frac{x_i - \mu}{\sigma} \right) \right] \\
& + (b-1) \sum_{i=1}^{n} \log \left\{ 1 - \exp\left(-a \exp\left[-\left(\frac{x_i - \mu}{\sigma} \right) \right] \right) \right\}.
\end{aligned}
$$

We determine the MLEs of the model parameters by setting these equations to zero and solving them simultaneously. We can use the `MaxBFGS` subroutine of the matrix programming language `Ox` (see, for example, Doornik (2007) [150]) and the `NLMixed` procedure in SAS to obtain $\widehat{\boldsymbol{\theta}}$

For interval estimation on the model parameters, we require the expected information matrix. The elements of the 4×4 observed information matrix $J(\boldsymbol{\theta})$ are given by $-\partial^2 \ell(\boldsymbol{\theta})/\partial\boldsymbol{\theta}\partial\boldsymbol{\theta}^{\top}$.

Under standard regularity conditions, the asymptotic distribution of $(\widehat{\boldsymbol{\theta}} - \boldsymbol{\theta})$ is $N_4\left(0, K(\boldsymbol{\theta})^{-1}\right)$, where $K(\boldsymbol{\theta}) = \mathrm{E}\{J(\boldsymbol{\theta})\}$ is the expected information matrix. If the expected information matrix does not have a closed-form, we can use $J(\widehat{\boldsymbol{\theta}})$ instead of $K(\widehat{\boldsymbol{\theta}})$. Then, the multivariate normal $N_4(0, J(\widehat{\boldsymbol{\theta}})^{-1})$ distribution can be used to construct approximate confidence intervals for the model parameters. The elements of the expected information matrix can be evaluated numerically.

The LR statistic can be used for testing goodness of fit of the KwGum distribution and for comparing this distribution with some of its special models.

6.3.6.   Bootstrap Re-sampling Methods

Efron (1979) [187] pioneered the bootstrap re-sampling approaches, where Q bootstrap samples with equal size are generated from the original sample. We can estimate some properties of the population based on these generated samples. The re-sampling technique has the non-parametric and parametric forms. We investigate the parametric bootstrap method, where the distribution function F is estimated by $\widehat{F}_{\boldsymbol{\theta}}$ from the bootstrap samples generated by the parametric model.

Let $(T_1, \ldots, T_n)$ be a set of observations from which an estimator $\widehat{\boldsymbol{\theta}} = s(\widehat{F})$ is determined from a parameter of interest $\boldsymbol{\theta} = s(F)$. Then, the $T_1^*, \ldots, T_Q^*$ samples are randomly generated through parametric bootstrap sampling. For the Q bootstrap generated samples, the bootstrap estimate of the parameter of interest for the qth sample is

$$\widehat{\boldsymbol{\theta}}_q^* = s\left(T_q^*\right),$$

i.e., it is the value of $\widehat{\boldsymbol{\theta}}$ for the sample T_q^*, $q = 1, \ldots, Q$.

The bootstrap estimator EP_Q of the standard error is defined as the standard deviation of these bootstrap samples (Efron and Tibshirani, 1994 [188])

$$\widehat{EP}_Q = \left[\frac{1}{(Q-1)} \sum_{q=1}^{Q} \left(\widehat{\theta}_q^* - \bar{\theta}_Q\right)^2\right]^{1/2},$$

where Q is the number of generated bootstrap samples and

$$\bar{\theta}_Q = Q^{-1} \sum_{q=1}^{Q} \widehat{\theta}_q^*.$$

Following Efron and Tibshirani (1994) [188], $Q \geq 200$ is quite sufficient to provide accurate bootstrap estimates. However, we have to take a high value for Q to obtain greater accuracy. Further, we explore the methods for the bias corrected and accelerated (BCa) to determine approximated confidence intervals from the bootstrap re-sampling approach. For more details, see Efron and Tibshirani (1994) [188] and Diciccio and Efron (1996) [189].

In the BCa method, the percentiles adopted to determine the bootstrap confidence intervals are functions of the corrections to tendency $\widehat{u}$ and acceleration $\widehat{z}_0$. The bias correction $\widehat{z}_0$ is calculated as the proportion of the estimates of bootstrap samples that are smaller than the original estimate $\widehat{\boldsymbol{\theta}}$, namely

where $\Phi^{-1}(\cdot)$ is the inverse of the standard normal cdf, $\widehat{\boldsymbol{\theta}}$ is the MLE from the observed sample and $\widehat{\boldsymbol{\theta}}_q^*$ from the MLE of the qth bootstrap sample.

Let $\widehat{\boldsymbol{\theta}}_{(i)}$ be the MLE of the sample without the ith observation. Then, $\widehat{u}$ is given by

$$\widehat{z}_0 = \Phi^{-1}\left(\frac{\sharp\left(\widehat{\boldsymbol{\theta}}_q^* < \widehat{\boldsymbol{\theta}}\right)}{Q}\right), \quad q = 1, \ldots, Q,$$

where $\Phi(\cdot)$ is the standard normal cdf, $\Phi^{-1}(\cdot)$ is its inverse, $\widehat{\boldsymbol{\theta}}$ is the MLE of the observed sample and $\widehat{\boldsymbol{\theta}}_q^*$ is the MLE of the qth bootstrap sample.

Let $\widehat{\boldsymbol{\theta}}_{(i)}$ be the MLE of the sample without the ith observation. Then, $\widehat{u}$ is given by

$$\widehat{u} = \frac{\sum_{i=1}^{n}\left[\widehat{\boldsymbol{\theta}}_{(\cdot)} - \widehat{\boldsymbol{\theta}}_{(i)}\right]^3}{6\left\{\sum_{i=1}^{n}\left[\widehat{\boldsymbol{\theta}}_{(\cdot)} - \widehat{\boldsymbol{\theta}}_{(i)}\right]^2\right\}^{3/2}}.$$

Note that $\widehat{\boldsymbol{\theta}}_{(\cdot)} = \sum_{i=1}^{n}\widehat{\boldsymbol{\theta}}_{(i)}/n$ and n is the sample size.

Thus, the BCa bootstrap interval of coverage $100(1 - 2\alpha)\%$ is

where

$$\alpha_1 = \Phi\left\{\widehat{z}_0 + \frac{\widehat{z}_0 + \Phi^{-1}(\alpha)}{1 - \widehat{u}\left[\widehat{z}_0 + \Phi^{-1}(\alpha)\right]}\right\}$$

and

$$\alpha_2 = \Phi\left\{\widehat{z}_0 + \frac{\widehat{z}_0 + \Phi^{-1}(1 - \alpha)}{1 - \widehat{u}\left[\widehat{z}_0 + \Phi^{-1}(1 - \alpha)\right]}\right\}.$$

The quantities α_1 and α_2 are simple corrections to the bootstrap percentiles with the other quantities defined before. The percentile bootstrap interval is considered as a special case of the BCa bootstrap interval (Efron and Tibshirani, 1994 [188]).

6.3.7. A Bayesian Analysis

In addition, we employ a Bayesian approach as an alternative source of analysis. The statistical knowledge obtained from the model parameters is achieved through the posterior marginal distribution. The Gibbs sampler and the Metropolis-Hasting algorithm are used to outperform analytical intractability.

In this context, we consider that the parameters a, b, μ and σ have independent priors, *i.e.*

$$\pi(a, b, \mu, \sigma) = \pi(a)\,\pi(b)\,\pi(\mu)\,\pi(\sigma),$$

where

- $a \sim \Gamma(a_1, b_1)$, a_1 and b_1 known;
- $b \sim \Gamma(a_2, b_2)$, a_2 and b_2 known;
- $\mu \sim N(\mu_0, \sigma_0^2)$, μ_0 and σ_0^2 known;
- $\sigma \sim \Gamma(a_3, b_3)$, a_3 and b_3 known,

and $\Gamma(a_i, b_i)$ denotes the gamma distribution with mean a_i/b_i and variance a_i/b_i^2 for $a_i > 0$ and $b_i > 0$.

We also consider independence among the parameters a, b, μ and σ. The joint posteriori distribution for a, b, μ and σ is

$$\pi(a, b, \sigma, \mu | x) \;\propto\; \left(\frac{a\,b}{\sigma}\right)^n \exp\left\{ \sum_{i=1}^{n} -\left(\frac{x_i - \mu}{\sigma}\right) - a\sum_{i=1}^{n} \exp\left[-\left(\frac{x_i - \mu}{\sigma}\right)\right] \right\}$$

$$\times \prod_{i=1}^{n}\left\{ 1 - \exp\left(-a\exp\left[-\left(\frac{x_i - \mu}{\sigma}\right)\right]\right) \right\}^{b-1} \pi(a, b, \sigma, \mu).$$

The conditional posteriori density functions are given by

$$\pi(a | x, b, \sigma, \mu) \;\propto\; a^n \exp\left\{ -a\sum_{i=1}^{n} \exp\left[-\left(\frac{x_i - \mu}{\sigma}\right)\right] \right\},$$

$$\times \prod_{i=1}^{n}\left\{ 1 - \exp\left(-a\exp\left[-\left(\frac{x_i - \mu}{\sigma}\right)\right]\right) \right\}^{b-1} \pi(a),$$

$$\pi(b | x, a, \sigma, \mu) \propto b^n \prod_{i=1}^{n}\left\{ 1 - \exp\left(-a\exp\left[-\left(\frac{x_i - \mu}{\sigma}\right)\right]\right) \right\}^{b} \pi(b),$$

$$\pi(\mu | x, a, b, \sigma) \;\propto\; \exp\left(\frac{n\mu}{\sigma}\right) \exp\left\{ -a\sum_{i=1}^{n} \exp\left[-\frac{(x_i - \mu)}{\sigma}\right] \right\}$$

$$\times \prod_{i=1}^{n}\left\{ 1 - \exp\left(-a\exp\left[-\left(\frac{x_i - \mu}{\sigma}\right)\right]\right) \right\}^{b-1} \pi(\mu)$$

and

$$\pi(\sigma|x,a,b,\mu) \;\propto\; \sigma^{-n}\exp\left\{\sum_{i=1}^{n}-\frac{x_i-\mu}{\sigma}-a\sum_{i=1}^{n}\exp\left[-\frac{x_i-\mu}{\sigma}\right]\right\}$$
$$\times\prod_{i=1}^{n}\left\{1-\exp\left(-a\exp\left[-\left(\frac{x_i-\mu}{\sigma}\right)\right]\right)\right\}^{b-1}\pi(\sigma).$$

We use the Metropolis-Hastings algorithm to generate the variables a, b, μ and σ associated with the corresponding conditional posteriori distributions.

6.3.8. Application: Minimum Flow Data

In this section, we model the minimum discharge of at least seven consecutive days ($Q_{7,10}$) in a period of 38 years of the Cuiabá River (Cuiabá, Mato Grosso, Brazil). These data were discussed by Andrade *et al.* (2007) [190]. We calculate $Q_{7,10}$ from a data collected in 38 years (January 1962 to October 1999) referring to lower flows of No 66260001 hydrological station, installed in the Cuiabá River.

The MLEs of the parameters of the KwGum distribution fitted to these data (their SEs are in parentheses) are obtained using the MaxBFGS subroutine in Ox. They are $\widehat{a} = 10.347$ (54.219), $\widehat{b} = 267.03$ (1037.2), $\widehat{\mu} = 6.4045$ (1032.5) and $\widehat{\sigma} = 193.64$ (128.15). These results are quite similar if we adopt the NLMixed routine in the SAS software.

Parametric Bootstrap Method

Following the discussion of Section 6.3.6, we adopt the parametric bootstrap method (with $B = 3,000$) to find the estimated bootstrap and the BCa confidence intervals given in Table 6.9. The estimates from both methods are very similar. The SEs are found by inverting the estimated observed information matrix. The bootstrapping is used to estimate the SEs based on 3,000 replicates for which the second derivatives of the log-likelihood do not have closed-form. Note that the SEs of the estimators are larger than the parametric bootstrap estimates.

Table 6.9:   Parametric bootstrap from the KwGum model fitted to the minimum flow data.

Parameter	Estimate	SE	95% C.I. BCa
a	10.4110	0.3760	(9.6870, 10.9280)
b	266.700	8.4640	(251.770, 278.790)
μ	6.4450	0.0510	(6.3840, 6.5010)
σ	193.2100	7.6450	(179.720, 204.910)

Bayesian Analysis

We consider the following independent priors to perform the Gibbs sampler:

$$a \sim \Gamma(0.01, 0.01), \ b \sim \Gamma(0.01, 0.01), \ \mu \sim N(0, 10) \ \text{ and } \ \sigma \sim \Gamma(0.01, 0.01),$$

for which we have a vague prior distribution. We generate two parallel independent runs of the Metropolis-Hasting algorithm chain (with size 30,000) for each parameter based on these prior densities. We delete the first 5,000 iterations to eliminate the effects of the initial values and avoid correlation problems. We consider spacing of size ten to have a sample of size 3,000 from each chain.

We control the convergence of the Gibbs sampler using the methods developed by Cowles and Carlin (1996) [184]. Following Gelman and Rubin (1992) [185], we adopt the between and within sequence information to determine the scale reduction $\hat{R}$. For all cases, these values are close to one, thus indicating that the chain converges.

Table 6.10 gives the posterior summaries for the parameters of the KwGum distribution. The posteriori means in this table agree with the MLEs given before. The estimates from the three methods are quite similar as expected.

Table 6.10: Posterior summary results for three models fitted to the minimum flow data. Here, SD, 2.5% and 97.5% denote the standard deviation and percentiles from the posterior distributions of the parameters, respectively.

KwGum	Mean	SD	2.5%	97.5%	$\hat{R}$
a	10.3509	0.5067	9.3671	11.3403	1.00009
b	266.6602	0.2990	266.0652	267.2341	1.00012
μ	6.4275	0.4975	5.4411	7.3984	1.00006
σ	193.5787	0.3985	192.7718	194.3406	1.00021
BIC	383.0671				

Gumbel	Mean	SD	2.5%	97.5%	$\hat{R}$
a	1	-	-	-	-
b	1	-	-	-	-
μ	91.0001	1.0072	89.0454	92.9859	1.00018
σ	35.9405	0.9954	34.0196	37.8121	1.00034
BIC	413.3961				

Beta-Gumbel	Mean	SD	2.5%	97.5%	$\hat{R}$
a	57.5162	1.0073	55.5468	59.4702	1.00007
b	48.5404	0.9966	46.5373	50.4344	0.99999
μ	-19.0049	0.9942	-21.0337	-17.1211	1.00008
σ	260.7411	0.9962	258.7195	262.6429	1.00025
BIC	402.0378				

They are listed in Table 6.12. Based on these values, the KwGum model fits the current data better than the other two models.

6.4. CONCLUSIONS

Some of the other most important distributions in the Kw-G family published so far are: the Kumaraswamy generalized gamma by Pascoa *et al.* (2011) [91], Kumaraswamy skew-normal by Kazemi *et al.* (2011) [190], Kumaraswamy log-logistic by Santana *et al.* (2012) [191], Kumaraswamy Birnbaum-Sanders by

Table 6.11: MLEs of the parameters for the three models fitted to minimum flow data.

Model	a	b	μ	σ	AIC	BIC	CAIC
KwGum	10.3452	266.66	6.4212	193.58	391.4	398.0	392.7
Gumbel	1	1	91.0029	35.9617	392.9	396.1	393.2
Beta Gumbel	57.4851	48.4798	-19.0725	260.74	392.4	399.0	393.6

Table 6.12: Goodness-of-fit statistics.

Distribution	W^*	A^*
KwGum	0.1681	2.5878
Gumbel	0.2683	3.1101
Beta Gumbel	0.2002	2.7336

Saulo *et al.* (2012) [192], Kumaraswamy generalized half-normal by Cordeiro *et al.* [193], Kumaraswamy modified Weibull by Cordeiro *et al.* [194], Kumaraswamy normal by Correa *et al.* (2012) [195], Kumaraswamy Pareto by Bourguignon *et al.* (2013) [196], Kumaraswamy generalized Pareto by Nadarajah and Eljabri (2013) [197], Kumaraswamy generalized linear failure rate by Elbatal (2013) [198], Kumaraswamy Lomax by Shams (2013) [199] and Kumaraswamy generalized Rayleigh by Gomes *et al.* (2014) [200]. In this chapter, we study in details three special models in the Kw-G family, namely: the Kumaraswamy Weibull, Kumaraswamy Burr XII and Kumaraswamy Gumbel distributions. We provide some structural properties for these distributions, general methods for estimating their model parameters, Monte Carlo simulations to verify the robustness of the estimates, and some applications to real data sets. The structural properties include linear representations, moments, quantile and generating functions, mean deviations, Lorenz and Bonferroni curves, reliability, order statistics and their moments. The applications of these special models reveal that the Kw-G family is a rather flexible mechanism for fitting a wide spectrum of real world data.

Chapter 7

The Gamma-G Family of Distributions

Abstract: This chapter presents the gamma generalized family of distributions proposed by Zofragos and Balakrishnan (2009). Several mathematical properties are provided such as   representations for gamma-G density and cumulative functions, some generalized moments, quantile and generating functions and entropies. A bivariate generalization is presented. An application is performed in order to illustrate empirically the usefulness of this family.

Keywords:: Gamma-G Model; GGum; GLL; GLN; GN; GW; Mean deviation; Moment; Order statistic.

7.1. INTRODUCTION

Zografos and Balakrishnan (2009) [202] pioneered a family of univariate continuous distributions generated by gamma random variables. Let $G(x)$ be any parent cdf for $x \in \mathbb{R}$. They defined the *gamma-G* family with pdf $f(x)$ and cdf $F(x)$ given by

$$f(x) = \frac{1}{\Gamma(a)} \left\{ -\log[1 - G(x)] \right\}^{a-1} g(x) \qquad (7.1)$$

and

$$F(x) = \frac{\gamma\left(a, -\log\left[1 - G(x)\right]\right)}{\Gamma(a)} = \frac{1}{\Gamma(a)} \int_0^{-\log[1-G(x)]} t^{a-1}\, \mathrm{e}^{-t} \mathrm{d}t, \qquad (7.2)$$

respectively, for $a > 0$, where $g(x) = \mathrm{d}\,G(x)/\mathrm{d}\,x$, $\Gamma(a)$ is the gamma function, and $\gamma(a, z)$ is the incomplete gamma function defined by (1.7) in Section 1.4.

Gauss M. Cordeiro, Rodrigo B. Silva & Abraão D. C. Nascimento

The hrf corresponding to (7.1) becomes

$$h(x) = \frac{\{-\log[1 - G(x)]\}^{a-1} g(x)}{\Gamma(a, -\log[1 - G(x)])}, \tag{7.3}$$

where $\Gamma(a, z) = \int_z^\infty t^{a-1} e^{-t} dt$ denotes the upper incomplete gamma function.

The gamma-G family has the same parameters of the parent G plus an extra shape parameter $a > 0$. Henceforth, if X is a random variable with pdf (7.1), we write $X \sim$ gamma-G(a). Every new gamma-G model can be determined from a given G distribution. Clearly, the G distribution is the basic exemplar of the gamma-G family when $a = 1$.

Zografos and Balakrishnan (2009) [201] presented several motivations for the gamma-G family: if $X_{1:1}, \dots, X_{1:n}$ are the order statistics from a sequence of independent random variables with common pdf $g(\cdot)$, then the pdf of the nth lower statistic is given by (7.1). Further, if Z is a gamma random variable with shape parameter $a > 0$ and unit scale parameter, then $X = F^{-1}(\exp(Z))$ has the pdf (7.1). Finally, if Z is a log-gamma random variable, then $X = F^{-1}(\exp\{-\exp(Z)\})$ has the pdf (7.1).

Recently, several mathematical properties of (7.1) and (7.2) were investigated by Nadarajah *et al.* (2015) [202]. Zografos and Balakrishnan (2009) [201] proposed expressions for moments associated with special gamma-G models (which hold only for natural a), a general expression for the Shannon entropy and a maximum entropy characterization.

7.2. SPECIAL GAMMA-G MODELS

The gamma-G family density function (7.1) furnishes for greater flexibility to describe tail points and, therefore, can be widely employed in many areas of engineering and biology. In this section, we present five special cases of this family. Models deduced from the Equation (7.1) can be analytically tractable when the cdf $G(x)$ and the pdf $g(x)$ have simple analytic expressions.

7.2.1. The Gamma-Weibull Distribution

Consider $G(x) = 1 - \exp\{-(\beta x)^\alpha\}$ to be the Weibull cdf with scale parameter $\beta > 0$ and shape parameter $\alpha > 0$, the gamma-Weibull (GW) density function (for $x > 0$) becomes

$$f_{\mathcal{GW}}(x) = \frac{\alpha \beta^{\alpha a}}{\Gamma(a)} x^{a\alpha - 1} \exp\{-(\beta x)^\alpha\}. \tag{7.4}$$

Equation (7.4) is important because it extends many distributions previously considered in the literature. In fact, it is identical to the generalized gamma (Stacy, 1962 [203]) distribution.

The Weibull distribution is a special case when $a = 1$ and the gamma distribution is another special case when $\alpha = 1$. The half-normal distribution corresponds to $a = 3$ and $\alpha = 2$. In addition, the log-normal distribution is a limiting special case when a goes to infinity.

The cdf and hrf corresponding to (7.4) are

$$F_{\mathcal{G}W}(x) = \frac{\gamma[a, (\beta\,x)^\alpha]}{\Gamma(a)}$$

and

$$h_{\mathcal{G}W}(x) = \frac{\alpha\,\beta^{a\,\alpha}\,x^{a\alpha-1}\exp\{-(\beta x)^\alpha\}}{\left\{\Gamma(a) - \gamma[a, (\beta\,x)^\alpha]\right\}},$$

respectively.

7.2.2.   The Gamma-Normal Distribution

The gamma-normal (GN) distribution is defined from (7.1) by taking $G(x)$ and $g(x)$ to be the cdf and pdf of the normal $N(\mu, \sigma^2)$ distribution. Its pdf is

$$f_{\mathcal{G}N}(x) = \frac{1}{\Gamma(a)}\left\{-\log\left[1 - \Phi\left(\frac{x-\mu}{\sigma}\right)\right]\right\}^{a-1}\phi\left(\frac{x-\mu}{\sigma}\right),$$

where $x \in \mathbb{R}$, $\mu \in \mathbb{R}$ is a location parameter, $\sigma > 0$ is a scale parameter, $a > 0$ is a shape parameter, and $\phi(\cdot)$ and $\Phi(\cdot)$ are the pdf and cdf of the standard normal distribution, respectively. For $\mu = 0$ and $\sigma = 1$, we obtain the standard GN distribution. Further, this distribution with $a = 1$ becomes the normal distribution.

7.2.3.   The Gamma-Gumbel Distribution

Consider the Gumbel distribution with location parameter $\mu \in \mathbb{R}$ and scale parameter $\sigma > 0$, where the pdf and cdf (for $x \in \mathbb{R}$) are

$$g(x) = \frac{1}{\sigma}\exp\left\{\left(\frac{x-\mu}{\sigma}\right) - \exp\left(\frac{x-\mu}{\sigma}\right)\right\}$$

and

$$G(x) = 1 - \exp\left\{-\exp\left(\frac{x-\mu}{\sigma}\right)\right\},$$

respectively. The mean and variance are equal to $\mu - \gamma\sigma$ and $\pi^2\sigma^2/6$, respectively, where γ is the Euler's constant ($\gamma \approx 0.57722$). Inserting these expressions into (7.1) yields the gamma-Gumbel (GGu) density function

$$f_{\mathcal{GG}u}(x) = \frac{1}{\sigma\Gamma(a)} \exp\left\{ (a-1)\left(\frac{x-\mu}{\sigma}\right) + \left(\frac{x-\mu}{\sigma}\right) - \exp\left(\frac{x-\mu}{\sigma}\right)\right\},$$

where $x, \mu \in \mathbb{R}$ and $a, \sigma > 0$. The Gumbel distribution corresponds to $a = 1$.

7.2.4.   The Gamma-lognormal Distribution

Let $G(x)$ be the log-normal distribution with cdf

$$G(x) = 1 - \Phi\left(\frac{-\log(x) + \mu}{\sigma}\right)$$

for $x > 0$, $\sigma > 0$ and $\mu \in \mathbb{R}$. The gamma-lognormal (GLN) pdf (for $x > 0$) reduces to

$$f_{\mathcal{GLN}}(x) = \frac{1}{\sqrt{2\pi}\,\sigma\,\Gamma(a)\,x} \exp\left\{ -\frac{1}{2}\left[\frac{\log(x)-\mu}{\sigma}\right]^2\right\}$$
$$\times \left\{ -\log\left[\Phi\left(\frac{-\log(x)+\mu}{\sigma}\right)\right]\right\}^{a-1}.$$

For $a = 1$, we obtain the log-normal distribution.

7.2.5.   The Gamma-log-logistic Distribution

The pdf and cdf of the log-logistic (LL) distribution are (for $x, \alpha, \beta > 0$)

$$g(x) = \frac{\beta}{\alpha^\beta}\, x^{\beta-1}\left[1 + \left(\frac{x}{\alpha}\right)\right]^{-2}$$

and

$$G(x) = 1 - \left[1 + \left(\frac{x}{\alpha}\right)^\beta\right]^{-1}.$$

Inserting these expressions into (7.1) gives the gamma-log-logistic (GLL) density function (for $x > 0$)

$$f_{\mathcal{GLL}}(x) = \frac{\beta}{\alpha^\beta\,\Gamma(a)}\, x^{\beta-1}\left[1 + \left(\frac{x}{\alpha}\right)^\beta\right]^{-2}\left\{ \log\left[1 + \left(\frac{x}{\alpha}\right)^\beta\right]\right\}^{a-1}.$$

The LL distribution is obtained when $a = 1$.

7.3.   LINEAR REPRESENTATIONS

Some useful linear combinations for (7.1) and (7.2) can be derived based on the concept of exponentiated distributions. For any parent cdf $G(x)$, we define a random variable having the exp-G distribution with power parameter $a > 0$ (see Chapter 2.1), say $X \sim \text{exp-G}(a)$, by the pdf $h_a(x) = a\,G(x)^{a-1}g(x)$ and cdf $H_a(x) = G(x)^a$.

For any real parameter c and $z \in (0, 1)$, the following formula holds

$$[-\log(1-z)]^c = z^c + \sum_{i=0}^{\infty} q_i(c)\, z^{i+c+1}, \qquad (7.5)$$

where $q_0(c) = c/2$, $q_1(c) = c\,(3c + 5)/24$, $q_2(c) = c\,(c^2 + 5c + 6)/48$, $q_3(c) = c\,(15c^3 + 150c^2 + 485c + 502)/5760$, *etc.* The proof is given in details by Flajonet and Odlyzko (1990) (see Theorem 3A, p. 227 [204]).

For a real parameter $a > 0$, using (7.5), equation (7.1) can be expressed as

$$f(x) = \sum_{k=0}^{\infty} b_k\, h_{a+k}(x), \qquad (7.6)$$

where $h_{a+k}(x)$ denotes the pdf of the exp-G$(a + k)$ distribution and the coefficients are: $b_0 = 1/\Gamma(a+1)$, $b_1 = q_0(a-1)/[(a+1)\Gamma(a)]$, $b_2 = q_1(a-1)/[(a+2)\Gamma(a)]$, $b_3 = q_2(a-1)/[(a+3)\Gamma(a)]$, *etc.*

Equation (7.6) reveals that the gamma-G family density is a linear combination of exp-G densities. Then, several properties of this family can be obtained by knowing those of the exp-G distribution, see, for example, Mudholkar *et al.* (1995) [36], Gupta and Kundu (2001) [43] and Nadarajah and Kotz (2006) [41], among others.

In a similar manner, the cdf (7.2) can be expressed as

$$F(x) = \sum_{k=0}^{\infty} b_k\, H_{a+k}(x), \qquad (7.7)$$

where $H_{a+k}(x)$ denotes the cdf of the exp-G$(a + k)$ distribution.

7.4.   ASYMPTOTES and SHAPES

The asymptotes of (7.1), (7.2) and (7.3) as $x \to -\infty, \infty$ are given by

$$f(x) \sim \frac{1}{\Gamma(a)} G^{a-1}(x)\, g(x) \qquad \text{as} \quad x \to -\infty,$$

$$f(x) \sim \frac{1}{\Gamma(a)} \left\{-\log[1 - G(x)]\right\}^{a-1} g(x) \quad \text{as} \quad x \to \infty,$$

$$F(x) \sim \frac{1}{\Gamma(a+1)} \left\{-\log[1 - G(x)]\right\}^a \quad \text{as} \quad x \to -\infty,$$

$$1 - F(x) \sim \frac{1}{\Gamma(a)} \left\{-\log[1 - G(x)]\right\}^{a-1} [1 - G(x)] \quad \text{as} \quad x \to \infty,$$

$$h(x) \sim \frac{1}{\Gamma(a)} G^{a-1}(x) \, g(x) \quad \text{as} \quad x \to -\infty,$$

and

$$f(x) \sim \frac{g(x)}{1 - G(x)} \quad \text{as} \quad x \to \infty.$$

So, the gamma-G density function behaves like the G hrf for very large x, whereas its hrf is proportional to the exp-G pdf for very small x.

The shapes of (7.1) and (7.3) can be described analytically. The critical points of the pdf are the roots of the equation:

$$\frac{g'(x)}{g(x)} = \frac{(1 - a) \, g(x)}{\log[1 - G(x)] \, [1 - G(x)]}. \tag{7.8}$$

There may be more than one root to (7.8). If $x = x_0$ is a root of (7.8) then it corresponds to a local maximum, a local minimum or a point of inflexion depending on whether $\lambda(x_0) < 0$, $\lambda(x_0) > 0$ or $\lambda(x_0) = 0$, where

$$\lambda(x) = \frac{(a - 1) \, g^2(x) \left\{2 - a + \log[1 - G(x)]\right\}}{\log^2[1 - G(x)] \, [1 - G(x)]^2}.$$

The critical points of the hrf are the roots of the equation:

$$\frac{g'(x)}{g(x)} = \frac{(1 - a) \, g(x)}{\log[1 - G(x)] \, [1 - G(x)]} - \frac{\left\{-\log[1 - G(x)]\right\}^{a-1} g(x)}{\Gamma\left(a, -\log[1 - G(x)]\right)}. \tag{7.9}$$

There may be more than one root to (7.9). If $x = x_0$ is a root of (7.9) then it corresponds to a local maximum, a local minimum or a point of inflexion depending on whether $\tau(x_0) < 0$, $\tau(x_0) > 0$ or $\tau(x_0) = 0$, where

$$\begin{aligned}
\tau(x) \quad &= \frac{(a - 1)g'(x)}{\log[1 - G(x)] \, [1 - G(x)]} + \frac{(a - 1)g^2(x) \left\{1 + \log[1 - G(x)]\right\}}{\log^2[1 - G(x)] \, [1 - G(x)]^2} \\
&+ \frac{g''(x)}{g(x)} - \left[\frac{g'(x)}{g(x)}\right]^2 - \frac{g^2(x) \left\{-\log[1 - G(x)]\right\}^{2a-2}}{\Gamma^2\left(a, -\log[1 - G(x)]\right)} \\
&+ \frac{\left\{-\log[1 - G(x)]\right\}^{a-2}}{\Gamma\left(a, -\log[1 - G(x)]\right)} \left\{\frac{(a - 1)g^2(x)}{1 - G(x)} - g'(x) \log[1 - G(x)]\right\}.
\end{aligned}$$

7.5. QUANTILE FUNCTION

Henceforth, we use an equation by Gradshteyn and Ryzhik (2000, Section 0.314 [11]) for a power series raised to a positive integer n

$$\left(\sum_{i=0}^{\infty} a_i\, u^i\right)^n = \sum_{i=0}^{\infty} c_{n,i}\, u^i, \tag{7.10}$$

where the coefficients $c_{n,i}$'s (for $i = 1, 2, \ldots$) can be determined from the recurrence equation

$$c_{n,i} = (i\, a_0)^{-1} \sum_{m=1}^{i} \left[m\,(n+1) - i\right] a_m\, c_{n,i-m}, \tag{7.11}$$

and $c_{n,0} = a_0^n$. The coefficient $c_{n,i}$ follows from $c_{n,0}, \ldots, c_{n,i-1}$ and hence from the quantities $a_0, \ldots, a_i$.

The generation of X is very easy: if V is a gamma random variable with shape parameter a and unit scale parameter, then

$$X = Q_G\left(1 - \exp\left[-V\right]\right)$$

will also be a gamma-G random variable. Further, inverting $F(x) = u$, we obtain (for $0 < u < 1$)

$$Q(u) = F^{-1}(u) = Q_G\left(1 - \exp\left[-Q^{-1}(a, 1 - u)\right]\right), \tag{7.12}$$

where $Q^{-1}(a, u)$ is the inverse function of $Q(a, x) = 1 - \gamma(a, x)/\Gamma(a)$, see `http://functions.wolfram.com/GammaBetaErf/InverseGammaRegularized/` for details. The asymptotes of (7.12) can be determined using known properties of $Q^{-1}(a, u)$. Using `http://functions.wolfram.com/GammaBetaErf/InverseGammaRegularized/06/02/01/`, one can see as $u \to 0$ that

$$Q(u) \sim Q_G\left(1 - \exp\left[-(1 - a)W_{-1}\left(-\frac{(1 - u)^{1/(a-1)}\Gamma 1/(a-1)(a)}{a - 1}\right)\right]\right),$$

where $W_{-1}(\cdot)$ denotes the product log function. Based on the site `http://functions.wolfram.com/GammaBetaErf/InverseGammaRegularized/06/01/03/0001/`, one can see as $u \to 1$ that

$$Q(u) \sim Q_G\left(1 - \exp\left[-(1 - u)^{1/a}\, \Gamma^{1/a}(a + 1) + \frac{(1 - u)^2\, \Gamma^{2/a}(a + 1)}{(a + 1)}\right]\right).$$

Quantiles of interest can be obtained from (7.12) by substituting appropriate values for u. Let $Q_G(u) = G^{-1}(u)$ be the baseline qf. In particular, the median of X is

$$\text{Median}(X) = Q_G\left(1 - \exp\left[-Q^{-1}(a, 1/2)\right]\right).$$

One can also use (7.12) for simulating gamma-G variates: if U is a uniform random variable on the unit interval $[0, 1]$ then

$$X = Q_G\left(1 - \exp\left[-Q^{-1}(a, 1 - U)\right]\right)$$

will be a gamma-G random variable.

If V is a gamma random variable with shape parameter a and unit scale parameter, the qf of V, say $Q_V(u)$, admits the power series

$$Q_V(u) = \sum_{i=0}^{\infty} m_i \left[\Gamma(a+1)\, u\right]^{i/a},$$

where $m_0 = 0$, $m_1 = 1$ and any coefficient m_{i+1} (for $i \geq 1$) is determined by the cubic recurrence equation

$$m_{i+1} = \frac{1}{i(a+i)}\left\{ \sum_{r=1}^{i} \sum_{s=1}^{i-s+1} s\,(i - r - s + 2)\, m_r\, m_s\, m_{i-r-s+2} \right.$$
$$\left. - \Delta(i) \sum_{r=2}^{i} r\,[r - a - (1-a)(i + 2 - r)]\, m_r m_{i-r+2} \right\},$$

where $\Delta(i) = 0$ if $i < 2$ and $\Delta(i) = 1$ if $i \geq 2$. The first few coefficients are $m_2 = 1/(a+1)$, $m_3 = (3a+5)/[2(a+1)^2(a+2)]$, ... We use the fact that $m_0 = 0$ and define (for $i = 0, 1, 2\ldots$) $t_i = m_{i+1}\, \Gamma(a+1)^{(i+1)/2}$ to express the gamma-G qf $Q(u) = Q_G(1 - \exp\{-Q_V(u)\})$ as

$$Q(u) = Q_G\left(1 - \sum_{k=0}^{\infty} \frac{(-1)^k}{k!} \left[\sum_{i=0}^{\infty} t_i\, u^{(i+1)/a}\right]^k\right).$$

Based on equations (7.10) and (7.11), we can write

$$Q(u) = Q_G\left(1 - \sum_{i,k=0}^{\infty} \frac{(-1)^k\, d_{k,i}}{k!}\, u^{(i+k)/a}\right),$$

where (for $k \geq 0$) $d_{k,0} = t_0^k$ and, for $i = 1, 2, \ldots$,

$$d_{k,i} = (i\, t_0)^{-1} \sum_{j=1}^{i} [j(k+1) - i]\, t_j\, d_{k,i-k}.$$

For $s \geq 1$, we define $J_s = \{(i,k); i + k = s, i, k = 1, 2, \ldots\}$ and $\nu_s = \sum_{(i,k) \in J_s}^{\infty} (-1)^{k+1} d_{k,i}/k!$. Then, we obtain

$$Q(u) = Q_G \left(\sum_{s=1}^{\infty} \nu_s\, u^{s/a} \right). \tag{7.13}$$

Hence, equation (7.13) indicates that the gamma-G qf can be formulated as the G qf used to a power series. This expansion holds for any gamma-G model. Thus, several moment-based measures of X can be simplified to integrals over $(0,1)$. For the great majority of these quantities, we can adopt ten terms in this power series.

Let $W(\cdot)$ be any integrable function on the real line. We can write

$$\int_{-\infty}^{\infty} W(x)\, f(x)\mathrm{d}x = \int_0^1 W\left(Q_G \left[\sum_{s=1}^{\infty} \nu_s\, u^{s/a} \right] \right) \mathrm{d}u. \tag{7.14}$$

Equations (7.13) and (7.14) are the main results of this section. In fact, several mathematical properties of X follow by using the right integral for special $W(\cdot)$ functions, which will be usually more simple than if they are based on the left integral.

For example, we consider a simple application for the nth moment of X. It follows from equations (7.10), (7.11) and (7.14)

$$E(X^n) = \sum_{s=1}^{\infty} \frac{f_{n,s}}{(s/a + 1)},$$

where $f_{n,0} = \nu_0^n$ and (for $s \geq 1$) $f_{n,s} = (s\,\nu_0)^{-1} \sum_{p=1}^{i} [p\,(n+1) - s]\, \nu_p\, f_{n,s-p}$.

7.6.   MOMENTS

Let $Y_k \sim$ exp-G$(a + k)$. A first formula for the nth moment of X can be obtained from (7.6) as

$$\mu_n' = \mathrm{E}(X^n) = \sum_{k=0}^{\infty} b_k\, E(Y_k^n). \tag{7.15}$$

Expressions for moments of several exponentiated distributions are given by Nadarajah and Kotz (2006) [41], which can be used to produce $\mathrm{E}(X^n)$. A second formula for $E(X^n)$ can be determined from (7.6) as

$$\mu_n' = \mathrm{E}(X^n) = \sum_{k=0}^{\infty} (a + k)\, b_k\, \tau(n, a + k - 1), \tag{7.16}$$

where

$$\tau(n,a) = \int_{-\infty}^{\infty} x^n \, G(x)^a \, g(x) dx = \int_0^1 Q_G(u)^n \, u^a du.$$

The ordinary moments of several gamma-G distributions can be determined directly from (7.16). Here, we give two examples. For the gamma-standard logistic, where $G(x) = (1+e^{-x})^{-1}$, using a result from Prudnikov *et al.* (1986, Section 2.6.13, equation 4), we obtain (for $t < 1$)

$$\mu'_n = \sum_{k=0}^{\infty} (a+k) \, b_k \left(\frac{\partial}{\partial t}\right)^n B(t+a+k, 1-t)\Big|_{t=0},$$

where $B(a,b)$ is the beta function (see (1.8) in Section 1.4). The moments of the gamma-exponential (with parameter $\lambda > 0$) are

$$\mu'_n = n! \, \lambda^n \sum_{k,j=0}^{\infty} \frac{(-1)^{n+j} \, (a+k) \, b_k}{(j+1)^{n+1}} \binom{a+k-1}{j}.$$

The central moments (μ_r), cumulants (κ_r) and factorial moments of X can be easily obtained from well-known relationships.

The nth incomplete moment of X follows from (7.6) as

$$m_n(y) = \int_{-\infty}^{y} x^n \, f(x) dx = \sum_{k=0}^{\infty} (a+k) \, b_k \int_0^{G(y)} Q_G(u)^n \, u^{a+k-1} du.$$

This integral can be evaluated numerically for most baseline G distributions.

7.7.   GENERATING FUNCTION

In this section, we provide two formulae for the mgf $M(t) = E[\exp(t\,X)]$ of X. Clearly, a first formula for $M(t)$ comes from (7.6) as

$$M(t) = \sum_{k=0}^{\infty} b_k \, M_k(t), \qquad (7.17)$$

where $M_k(t)$ is the mgf of Y_k. Hence, $M(t)$ can be obtained from the generating function of the exp-G distribution.

A second formula for $M(t)$ can be derived from (7.6) as

$$M(t) = \sum_{i=0}^{\infty} (a+k) \, b_k \, \rho(t, a+k-1), \qquad (7.18)$$

where the integral

$$\rho(t,a) = \int_{-\infty}^{\infty} \exp(tx)\, G(x)^a\, g(x)dx = \int_0^1 \exp\left\{t\, Q_G(u)\right\} u^a du$$

can be evaluated from $Q_G(u)$ at least numerically.

We can determine the mgfs of several gamma-G distributions directly from equation (7.18). For example, the mgfs of the gamma-exponential (with parameter λ and $t < \lambda^{-1}$) and gamma-standard logistic (for $t < 1$) take the forms

$$M(t) = \sum_{k=0}^{\infty} (a+k)\, b_k\, B(a+k, 1-\lambda t)$$

and

$$M(t) = \sum_{k=0}^{\infty} (a+k)\, b_k\, B(t+a+k, 1-t),$$

respectively.

7.8. MEAN DEVIATIONS

The mean deviations about the mean ($\delta_1 = E(|X-\mu_1'|)$) and about the median ($\delta_2 = E(|X - M|)$) of X can be expressed as

$$\delta_1 = 2\mu_1'\, F(\mu_1') - 2m_1(\mu_1') \qquad \text{and} \qquad \delta_2 = \mu_1' - 2m_1(M), \qquad (7.19)$$

respectively, where $\mu_1' = E(X)$, M is the median given by (7.12) at $u = 0.5$, $F(\mu_1')$ is easily evaluated from the cdf in (7.2) and $m_1(z) = \int_{-\infty}^z x f(x)dx$ is the first incomplete moment.

In this section, we provide two alternative ways to compute δ_1 and δ_2. A general equation for $m_1(z)$ can be derived from (7.6) as

$$m_1(z) = \sum_{k=0}^{\infty} b_k\, J_k(z), \qquad (7.20)$$

where

$$J_k(z) = \int_{-\infty}^z x\, h_{a+k}(x)\, \mathrm{d}x. \qquad (7.21)$$

Equation (7.21) is the basic quantity to compute the mean deviations of the exp-G distributions. Hence, the mean deviations in (7.19) depend only on the mean deviations of the exp-G distribution. So, alternative representations for δ_1 and δ_2 are

$$\delta_1 = 2\mu_1' F\left(\mu_1'\right) - 2\sum_{k=0}^{\infty} b_k J_k\left(\mu_1'\right) \qquad \text{and} \qquad \delta_2 = \mu_1' - 2\sum_{k=0}^{\infty} b_k J_k(M).$$

A second general formula for $m_1(z)$ can be derived by setting $u = G(x)$ in equation (7.6)

$$m_1(z) = \sum_{k=0}^{\infty} (a+k)\, b_k\, T_k(z), \qquad (7.22)$$

where

$$T_k(z) = \int_0^{G(z)} Q_G(u)\, u^{a+k-1} du. \qquad (7.23)$$

In a similar manner, the mean deviations of any gamma-G distribution can be determined from equations (7.22) and (7.23). For example, the mean deviations of the gamma-exponential (with parameter λ) and gamma-standard logistic follow by applying the generalized binomial theorem from the functions

$$T_k(z) = \lambda^{-1}\,\Gamma(a+k-1) \sum_{j=0}^{\infty} \frac{(-1)^j\,\{1 - \exp\left(-j\lambda z\right)\}}{\Gamma(a+k-1-j)\,(j+1)!}$$

and

$$T_k(z) = \frac{1}{\Gamma(k)} \sum_{j=0}^{\infty} \frac{(-1)^j\,\Gamma(a+k+j)\,\{1 - \exp(-jz)\}}{(j+1)!},$$

respectively.

Applications of these equations can be addressed to obtain Bonferroni and Lorenz curves defined (for a probability π) by $B(\pi) = T(q)/(\pi\mu_1')$ and $L(\pi) = T(q)/\mu_1'$, respectively, where $\mu_1' = E(X)$ and $q = Q_G(1 - \exp[-Q^{-1}(a, 1-\pi)])$ is the qf given by (7.12) evaluated at π.

7.9. ENTROPIES

The Rényi entropy of a random variable with pdf $f(\cdot)$ is defined by

$$I_R(\gamma) = \frac{1}{1-\gamma} \log\left(\int_0^{\infty} f(x)^\gamma \mathrm{d}x\right)$$

for $\gamma > 0$ and $\gamma \neq 1$. The Shannon entropy of a random variable X is defined by $E[-\log f(X)]$. It is the particular case of the Rényi entropy for $\gamma \uparrow 1$. The generalized binomial coefficient to real arguments is given by $\binom{x}{y} = \Gamma(x+1)/[\Gamma(y+1)\Gamma(x-y+1)]$. For any real parameter $a > 0$, the following formula holds

(http://functions.wolfram.com/ElementaryFunctions/Log/06/01/04/03/)

$$
\begin{aligned}
\{-\log[1-G(x)]\}^{a-1} &= (a-1)\sum_{k=0}^{\infty}\binom{k+1-a}{k} \\
&\times \sum_{j=0}^{k}\frac{(-1)^{j+k}\,p_{j,k}}{(a-1-j)}\binom{k}{j}G(x)^{a+k-1}, \quad (7.24)
\end{aligned}
$$

where the constants $p_{j,k}$ can be determined recursively by (for $j \geq 0, k \geq 1$)

$$
p_{j,k} = \frac{1}{k}\sum_{m=1}^{k}\frac{(-1)^m\,[m(j+1)-k]}{(m+1)}p_{j,k-m} \quad (7.25)
$$

and $p_{j,0} = 1$.

Next, we derive expressions for the Rényi and Shannon entropies when X is a gamma-G random variable. By using (7.24), we can write

$$
\begin{aligned}
\{-\log[1-G(x)]\}^{\gamma(a-1)} &= \gamma(a-1)\sum_{k=0}^{\infty}\binom{k-\gamma(a-1)}{k} \\
&\times \sum_{j=0}^{k}\frac{(-1)^{j+k}\,p_{j,k}}{[\gamma(a-1)-j]}\binom{k}{j}G(x)^{\gamma(a-1)+k}.
\end{aligned}
$$

Then,

$$
\begin{aligned}
\int_0^{\infty}g(x)^{\gamma}\mathrm{d}x &= \int_0^{\infty}\frac{1}{\Gamma(a)^{\gamma}}\{-\log[1-G(x)]\}^{\gamma(a-1)}\,g(x)^{\gamma}\mathrm{d}x \\
&= \frac{1}{\Gamma(a)^{\gamma}}\sum_{k=0}^{\infty}\binom{k-\gamma(a-1)}{k}\sum_{j=0}^{k}\frac{(-1)^{j+k}\,p_{j,k}}{[\gamma(a-1)-j]}\binom{k}{j}I_k,
\end{aligned}
$$

where I_k is

$$
I_k = \int_0^{\infty}G(x)^{[\gamma(a-1)+k]}\,g(x)^{\gamma}\mathrm{d}x.
$$

Hence, the Rényi entropy of X reduces to

$$I_R(\gamma) = -\frac{\gamma \log \Gamma(a)}{1-\gamma} + \frac{1}{1-\gamma} \log \left\{ \sum_{k=0}^{\infty} \binom{k-\gamma a + \gamma}{k} \right.$$

$$\left. \times \sum_{j=0}^{k} \frac{(-1)^{j+k} p_{j,k}}{[\gamma(a-1)-j]} \binom{k}{j} I_k \right\}. \qquad (7.26)$$

The Shannon entropy can follow by limiting $\gamma \uparrow 1$ in (7.26). However, it is easier to obtain an expression for it from first principles. Using the power series for $\log(1-z)$, we can write

$$\begin{aligned}
\mathrm{E}[-\log f(X)] \\
&= \log \Gamma(a) + (1-a)\mathrm{E}\left[\log\{-\log[1-G(X)]\}\right] - \mathrm{E}\left[\log g(X)\right] \\
&= \log \Gamma(a) + (1-a)\mathrm{E}\left[\log\left\{ \sum_{r=1}^{\infty} \frac{G(X)^r}{r} \right\}\right] - \mathrm{E}\left[\log g(X)\right] \\
&= \log \Gamma(a) + (1-a)\mathrm{E}\left[\log\left\{ 1 + \sum_{r=2}^{\infty} \frac{G(X)^{r-1}}{r} \right\}\right] \\
&\quad + \mathrm{E}\left[\log G(X)\right] - \mathrm{E}\left[\log g(X)\right] \\
&= \log \Gamma(a) + (1-a)\left[\sum_{j=1}^{\infty} \frac{(-1)^{j-1}}{j} \mathrm{E}\left\{ G(X) \sum_{r=0}^{\infty} \frac{G(X)^r}{(r+2)} \right\}^j \right] \\
&\quad + \mathrm{E}\left[\log G(X)\right] - \mathrm{E}\left[\log g(X)\right].
\end{aligned}$$

For $j, r \geq 1$, we obtain from equations (7.10)–(7.11)

$$\left\{ G(X) \sum_{r=0}^{\infty} \frac{G(X)^r}{(r+2)} \right\}^j = \sum_{k=0}^{\infty} e_{j,r} \, G(X)^{r+j},$$

where $e_{j,0} = 2^{-j}$ and, for $r \geq 1$,

$$e_{j,r} = \frac{2}{r} \sum_{m=1}^{r} \frac{[m(j+1)-r]}{m+2} e_{r,r-m}.$$

Then,

$$\begin{aligned}
\mathrm{E}\left[-\log f(X)\right] &= \log \Gamma(a) + (1-a) \sum_{j=1}^{\infty} \frac{(-1)^{j-1}}{j} \sum_{r=0}^{\infty} e_{j,r} \, \mathrm{E}\left[G(X)^{r+j}\right] \\
&\quad + \mathrm{E}\left[\log G(X)\right] - \mathrm{E}\left[\log g(X)\right]. \qquad (7.27)
\end{aligned}$$

The three expectations in (7.27) can be evaluated numerically for given $G(\cdot)$ and $g(\cdot)$. By using the linear representation (7.6), they can be expressed as

$$\mathrm{E}\left[G(X)^{r+j}\right] = \sum_{k=0}^{\infty}(a+k)\,b_k \int_0^{\infty} G(x)^{a+r+j+k-1}\,g(x)\mathrm{d}x$$

$$= \sum_{k=0}^{\infty}\frac{(a+k)\,b_k}{(a+r+j+k)},$$

$$\mathrm{E}\left[\log G(X)\right] = \sum_{k=0}^{\infty}(a+k)\,b_k \int_0^{\infty} \log[G(x)]\,G(x)^{a+k-1}g(x)\mathrm{d}x$$

$$= -\sum_{k=0}^{\infty}\frac{b_k}{a+k}$$

and

$$\mathrm{E}\left[\log g(X)\right] = \sum_{k=0}^{\infty}(a+k)\,b_k \int_0^{\infty} \log[g(x)]\,G(x)^{a+k-1}g(x)\mathrm{d}x,$$

respectively. The last equation can also be given in terms of the baseline qf as

$$\mathrm{E}\left[\log g(X)\right] = \sum_{k=0}^{\infty}(a+k)\,b_k \int_0^{1} \log\left[g\left(Q_G(u)\right)\right]u^{a+k-1}\mathrm{d}u,$$

where this integral can be evaluated for most baseline distributions at least numerically.

7.10.   ORDER STATISTICS

Suppose $X_1,\ldots,X_n$ is a random sample from the gamma-G family. Let $X_{i:n}$ denote the ith order statistic. Based on equations (7.6) and (7.7), the pdf of $X_{i:n}$ can be expressed as

$$f_{i:n}(x) = K\,f(x)\,F(x)^{i-1}\left\{1 - F(x)\right\}^{n-i}$$

$$= K\sum_{j=0}^{n-i}(-1)^j\binom{n-i}{j}f(x)\,F(x)^{j+i-1}$$

$$= K\sum_{j=0}^{n-i}(-1)^j\binom{n-i}{j}\left[\sum_{r=0}^{\infty}b_r\,(a+r)\,G(x)^{a+r-1}\,g(x)\right]$$

$$\times \left[\sum_{k=0}^{\infty}b_k\,G(x)^{a+k}\right]^{j+i-1},$$

where $K = n!/[(i-1)!\,(n-i)!]$. Using (7.10) and (7.11), we can write

$$\left[\sum_{k=0}^{\infty} b_k\,G(x)^{a+k}\right]^{j+i-1} = \sum_{k=0}^{\infty} f_{j+i-1,k}\,G(x)^{a(j+i-1)+k},$$

where $f_{j+i-1,0} = b_0^{j+i-1}$ and (for $k \geq 1$)

$$f_{j+i-1,k} = (k\,b_0)^{-1}\sum_{m=1}^{k}[m(j+i)-k]\,b_m\,f_{j+i-1,k-m}.$$

Hence,

$$f_{i:n}(x) = \sum_{j=0}^{n-i}\sum_{r,k=0}^{\infty} m_{j,r,k}\,h_{(j+i)a+r+k}(x), \qquad (7.28)$$

where

$$m_{j,r,k} = \frac{(-1)^j\,n!}{(i-1)!\,(n-i-j)!\,j!}\,\frac{(a+r)\,b_r\,f_{j+i-1,k}}{[a(j+i)+r+k]}.$$

Equation (7.28) is the main result of this section. It provides the pdf of the gamma-G order statistics as a triple linear combination of exp-G density functions. Then, several mathematical quantities of these order statistics can be obtained from those gamma-G quantities.

7.11. LIKELIHOOD ESTIMATION

The estimation of the unknown parameters of the gamma-G family is investigated by the method of maximum likelihood. Let $x_1, \ldots, x_n$ be a sample from (7.1). Let $\boldsymbol{\theta}$ be a q-parameter vector specifying $G(\cdot)$. The log-likelihood function for the model parameters is given by

$$\ell(a,\boldsymbol{\theta}) = \sum_{i=1}^{n}\log g\,(x_i) + (a-1)\sum_{i=1}^{n}\log\{-\log[1-G\,(x_i)]\} - n\log\Gamma(a).$$

$$(7.29)$$

The first derivatives of ℓ with respect to the parameters a and $\boldsymbol{\theta}$ are:

$$\frac{\partial\ell}{\partial a} = \sum_{i=1}^{n}\log\{-\log[1-G\,(x_i)]\} - n\psi(a),$$

$$\frac{\partial\ell}{\partial\boldsymbol{\theta}} = \sum_{i=1}^{n}\frac{\partial g\,(x_i)\,/\partial\boldsymbol{\theta}}{g\,(x_i)} + (a-1)\sum_{i=1}^{n}\frac{\partial G\,(x_i)\,/\partial\boldsymbol{\theta}}{\log\{1-G\,(x_i)\}\{1-G\,(x_i)\}},$$

where $\psi(a) = d \log \Gamma(a)/da$ is the digamma function. The MLEs of $(a, \boldsymbol{\theta})$, say $(\widehat{a}, \widehat{\boldsymbol{\theta}})$, are the simultaneous solutions of the equations $\partial\ell/\partial a = 0$ and $\partial\ell/\partial\boldsymbol{\theta} = \mathbf{0}$.

Maximization of (7.29) can be performed using well-known routines like `nlm` or `optimize` in the R statistical package. Our numerical calculations prove that the surface of (7.29) is smooth for given smooth functions $g(\cdot)$ and $G(\cdot)$. The routines are able to locate the maximum in all cases and for different starting values. However, we perform the computations from reasonable starting values. These can be obtained, for example, by the method of moments. For $r = 1, \ldots, q+1$, let $m_r = n^{-1} \sum_{i=1}^{n} x_i^r$ denote the first $q+1$ sample moments.

Equating these moments to the theoretical versions given in Section 7.6, we have $m_r = \mathrm{E}(X^r)$, for $r = 1, \ldots, q+1$. These equations can be solved numerically to obtain the moments estimates.

For interval estimation of $(a, \boldsymbol{\theta})$ and hypotheses tests, one requires the observed information matrix for $(a, \boldsymbol{\theta})$ given by

$$\mathbf{J} = \begin{pmatrix} J_{11} & \mathbf{J}_{12} \\ \mathbf{J}_{12} & \mathbf{J}_{22} \end{pmatrix},$$

where

$$J_{11} = \frac{\partial^2 \ell}{\partial a^2} = n\psi'(a),$$

$$\mathbf{J}_{12} = \frac{\partial^2 \ell}{\partial a \partial \boldsymbol{\theta}} = -\sum_{i=1}^{n} \frac{\partial G(x_i)/\partial \boldsymbol{\theta}}{\log\{1 - G(x_i)\}\{1 - G(x_i)\}},$$

$$\mathbf{J}_{22} = \frac{\partial^2 \ell}{\partial \boldsymbol{\theta} \partial \boldsymbol{\theta}^\top} = -\sum_{i=1}^{n} \frac{\partial^2 g(x_i)/\partial \boldsymbol{\theta} \partial \boldsymbol{\theta}^\top}{g(x_l)} + \sum_{i=1}^{n} \frac{[\partial g(x_i)/\partial \boldsymbol{\theta}][\partial g(x_i)/\partial \boldsymbol{\theta}]^\top}{g^2(x_l)}$$

$$+ (1-a) \sum_{i=1}^{n} \frac{\partial^2 G(x_i)/\partial \boldsymbol{\theta} \partial \boldsymbol{\theta}^\top}{\log\{1 - G(x_i)\}\{1 - G(x_i)\}}$$

$$+ (1-a) \sum_{i=1}^{n} \frac{[\partial G(x_i)/\partial \boldsymbol{\theta}][\partial G(x_i)/\partial \boldsymbol{\theta}]^\top [1 + \log\{1 - G(x_i)\}]}{\log^2[1 - G(x_i)][1 - G(x_i)]^2}.$$

In large samples, the distribution of $(\widehat{a} - a, \widehat{\boldsymbol{\theta}} - \boldsymbol{\theta})$ can be approximated by a $(q+1)$ normal distribution with zero means and estimated variance-covariance matrix $\widehat{\mathbf{J}}^{-1}$. Some asymptotic properties of $(\widehat{a}, \widehat{\boldsymbol{\theta}})$ can be based on this normal approximation.

7.12. A BIVARIATE GENERALIZATION

An immediate bivariate generalization of (7.2) has the joint cdf

$$F(x, y) = \frac{\gamma\left(a, -\log\left[1 - G(x, y)\right]\right)}{\Gamma(a)} \tag{7.30}$$

for $x > 0$, $y > 0$ and $a > 0$. The joint pdf is

$$f(x, y) = \frac{1}{\Gamma(a)} \left\{-\log\left[1 - G(x, y)\right]\right\}^{a-2} A(x, y), \tag{7.31}$$

where

$$A(x, y) = \frac{a - 1}{1 - G(x, y)} \frac{\partial G(x, y)}{\partial x} \frac{\partial G(x, y)}{\partial y} - \log\left[1 - G(x, y)\right] \frac{\partial^2 G(x, y)}{\partial x \partial y}.$$

The marginal cdfs are

$$F(x) = \frac{\gamma\left(a, -\log\left[1 - G(x)\right]\right)}{\Gamma(a)}$$

and

$$F(y) = \frac{\gamma\left(a, -\log\left[1 - G(y)\right]\right)}{\Gamma(a)}.$$

The marginal pdfs are

$$f(x) = \frac{1}{\Gamma(a)} \left\{-\log[1 - G(x)]\right\}^{a-1} g(x)$$

and

$$f(y) = \frac{1}{\Gamma(a)} \left\{-\log[1 - G(y)]\right\}^{a-1} g(y).$$

The conditional cdfs are

$$F(x|y) = \frac{\gamma\left(a, -\log\left[1 - G(x, y)\right]\right)}{\gamma\left(a, -\log\left[1 - G(y)\right]\right)}$$

and

$$F(y|x) = \frac{\gamma\left(a, -\log\left[1 - G(x, y)\right]\right)}{\gamma\left(a, -\log\left[1 - G(x)\right]\right)}.$$

The conditional pdfs are

$$f(x|y) = \frac{\{-\log[1-G(x,y)]\}^{a-2} A(x,y)}{\{-\log[1-G(y)]\}^{a-1} g(y)}$$

and

$$f(y|x) = \frac{\{-\log[1-G(x,y)]\}^{a-2} A(x,y)}{\{-\log[1-G(x)]\}^{a-1} g(x)}.$$

Also,

$$\frac{\partial F(x,y)}{\partial x} = \frac{1}{\Gamma(a)} \{-\log[1-G(x,y)]\}^{a-1} \frac{\partial G(x,y)}{\partial x}$$

and

$$\frac{\partial F(x,y)}{\partial y} = \frac{1}{\Gamma(a)} \{-\log[1-G(x,y)]\}^{a-1} \frac{\partial G(x,y)}{\partial y}.$$

As in Section 7.3, it is useful to obtain power series for (7.30) and (7.31). First, we consider bivariate generalizations of the exp-G distribution

$$h_a(x,y) = a\, G(x,y)^{a-1} \frac{\partial^2 G(x,y)}{\partial x \partial y}$$
$$+ a(a-1)\, G(x,y)^{a-2} \frac{\partial G(x,y)}{\partial x} \frac{\partial G(x,y)}{\partial y}$$

and

$$H_a(x,y) = G(x,y)^a,$$

respectively.   Using the Taylor series expansion for the incomplete gamma function (see `http://functions.wolfram.com/GammaBetaErf/Gamma2/06/01/04/01/01/0003/`) and (7.24), equation (7.30) can be rewritten as

$$F(x,y) = \sum_{k,j=0}^{\infty} (a+k)\, \omega_{j,k}\, H_{a+j+k}(x,y),$$

where

$$\omega_{j,k} = \binom{j-k-a}{j} \sum_{i=0}^{j} \frac{(-1)^{i+j}\, p_{i,j}}{(a+k-i)} \binom{j}{i},$$

and the $p_{i,i}$'s are given by (7.25). Correspondingly, we can express (7.31) as

$$f(x, y) = \sum_{j,k=0}^{\infty} (a + k)\, \omega_{j,k}\, h_{a+j+k}(x, y).$$

This representation can be used to derive expressions for the joint mgf, joint cgf, product moments and other properties for the bivariate distribution (7.30).

We consider by maximum likelihood method to estimate the unknown parameters in (7.30). Suppose $G(\cdot, \cdot)$ and $g(\cdot, \cdot)$ are parameterized by $\boldsymbol{\theta}$. Let $(x_1, y_1), \ldots, (x_n, y_n)$ be a sample from (7.30). Then the log-likelihood function for a and $\boldsymbol{\theta}$ takes the form

$$\ell = \ell(a, \boldsymbol{\theta}) = \sum_{i=1}^{n} \log A(x_i, y_i) + (a - 2) \sum_{i=1}^{n} \log\{-\log[1 - G(x_i, y_i)]\}$$
$$- n \log \Gamma(a).$$

The first derivatives with respect to the parameters a and $\boldsymbol{\theta}$ are:

$$\frac{\partial \ell}{\partial a} = \sum_{i=1}^{n} \log\{-\log[1 - G(x_i, y_i)]\} - n\psi(a) + \sum_{i=1}^{n} \frac{\partial A(x_i, y_i)/\partial a}{A(x_i, y_i)},$$

and

$$\frac{\partial \ell}{\partial \boldsymbol{\theta}} = (a - 2) \sum_{i=1}^{n} \frac{\partial G(x_i, y_i)/\partial \boldsymbol{\theta}}{\log\{1 - G(x_i, y_i)\}\{1 - G(x_i, y_i)\}}$$
$$+ \sum_{i=1}^{n} \frac{\partial A(x_i, y_i)/\partial \boldsymbol{\theta}}{A(x_i, y_i)},$$

where

$$\frac{\partial A(x, y)}{\partial a} = \frac{1}{[1 - G(x, y)]} \frac{\partial G(x, y)}{\partial x} \frac{\partial G(x, y)}{\partial y},$$

and

$$\frac{\partial A(x, y)}{\partial \boldsymbol{\theta}} = \frac{(a - 1)}{[1 - G(x, y)]} \left\{ \frac{1}{[1 - G(x, y)]} \frac{\partial G(x, y)}{\partial \boldsymbol{\theta}} \frac{\partial G(x, y)}{\partial x} \frac{\partial G(x, y)}{\partial y} \right.$$
$$\left. + \frac{\partial G(x, y)}{\partial x \partial \boldsymbol{\theta}} \frac{\partial G(x, y)}{\partial y} + \frac{\partial G(x, y)}{\partial y \partial \boldsymbol{\theta}} \frac{\partial G(x, y)}{\partial x} \right\}$$
$$+ \frac{1}{[1 - G(x, y)]} \frac{\partial G(x, y)}{\partial \boldsymbol{\theta}} \frac{\partial^2 G(x, y)}{\partial x \partial y} - \log[1 - G(x, y)] \frac{\partial^3 G(x, y)}{\partial x \partial y \partial \boldsymbol{\theta}}.$$

The MLEs $(\widehat{a}, \widehat{\boldsymbol{\theta}})$ are the simultaneous solutions of the equations $\partial \ell / \partial a = 0$ and $\partial \ell / \partial \boldsymbol{\theta} = \mathbf{0}$.

For interval estimation of $(a, \boldsymbol{\theta})$ and tests of hypotheses, one requires the joint observed information matrix for $(a, \boldsymbol{\theta})$ given by

$$\mathbf{J} = \begin{pmatrix} J_{11} & \mathbf{J}_{12} \\ \mathbf{J}_{12} & \mathbf{J}_{22} \end{pmatrix}.$$

The elements of $\mathbf{J}$ can be computed numerically using standard analytical software.

7.13.   APPLICATION

We compare the fits of special gamma-G distributions, *i.e.*, the GW, GLN, GLL and Weibull distributions by means of a real data set. The model parameters are estimated by maximum likelihood (Section 7.11) using the **NLMixed** subroutine in **SAS**. The data on the failure times of 20 mechanical components given in Murthy *et al.* (2004) [57] are:

66.10, 79.00, 91.10, 94.90, 106.90, 112.80, 114.60, 131.20, 138.50, 144.50, 153.10, 159.30, 216.40, 229.70, 263.60, 266.20, 270.60, 275.70, 322.00, 544.60.

Table 7.1 lists the MLEs of the parameters (SEs given in parentheses) and the values of the AIC, CAIC and BIC statistics.   The figures in this table indicate that the GLL model provides the best fit to these data. Further, the GW and GLN models are much better than the Weibull model.

More information is provided by a visual comparison of the fitted density functions and the histogram of the data. The plots of the fitted GW, GLL, GLN and Weibull density functions and estimated cumulative functions are displayed in Figure 7.1. These plots indicate that the generated distributions provide adequate fits.

7.14.   THE RISTIĆ AND BALAKRISHNAN FAMILY

An alternative gamma-G family was proposed by Ristić and Balakrishnan (2012) [205], so-called the *Ristić-Balakrishnan-G* ("RB-G") family, for $x \in \mathbb{R}$

Table 7.1: MLEs of the model parameters for the mechanical components data, the corresponding SEs (given in parentheses) and the AIC, CAIC and BIC statistics.

Model	a	α	β	AIC	CAIC	BIC
GW model	19.2862	0.4247	6.1249	241.8	243.8	244.8
	(1.0304)	(0.0243)	(2.6140)			
GLL model	9.9057	30.0929	5.8567	240.5	242.0	243.4
	(3.6901)	(4.2582)	(1.3961)			
Weibull	1	1.8476	0.0047	243.4	244.1	245.4
	-	(0.2973)	(0.0006)			
	a	μ	σ			
GLN model	4.1067	3.7917	0.6479	240.9	242.4	243.9
	(2.4185)	(4.6695)	(0.2489)			

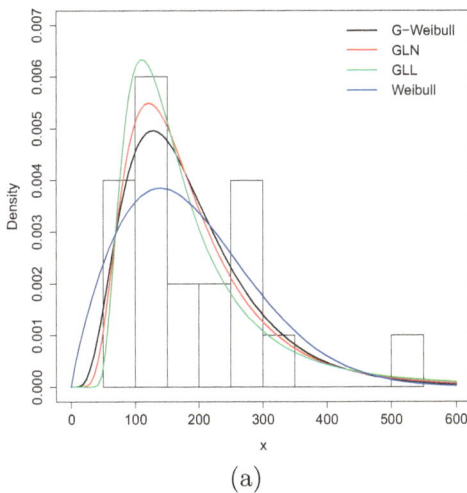

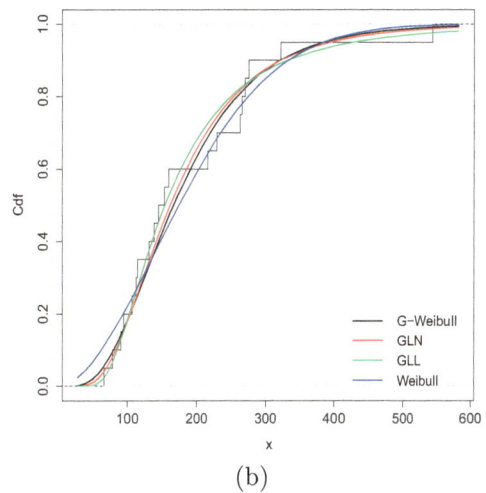

Figure 7.1: (a) Fitted GW, GLN, GLL and Weibull densities. (b) Estimated cumulative functions and the empirical cumulative distribution.

and $\alpha > 0$, having pdf and cdf given by

$$f(x) = \frac{g(x)}{\Gamma(\alpha)} \left\{-\log[G(x)]\right\}^{\alpha-1} \text{ and } F(x) = 1 - \frac{\gamma(\alpha, -\log[G(x)])}{\Gamma(\alpha)}, \quad (7.32)$$

respectively.

The RB-G family has the same parameters of the parent G distribution plus an extra positive shape parameter α. Each RB-G distribution can be obtained from a specified G model. For $\alpha = 1$, the parent G model is just a basic exemplar of the RB-G family.

For the rest of this chapter, we denote by X a random variable with pdf and cdf (7.32) and write $X \sim$RB-G(α). If the support of X is $(0, \infty)$, its hrf reduces to

$$h(x) = \frac{\{-\log[G(x)]\}^{\alpha-1} g(x)}{\gamma\left(\alpha, -\log[G(x)]\right)}.$$

The simulation of X is very easy. Let $Q_G(\cdot)$ be the baseline qf. By inverting $F(x) = u$ in (7.32), we obtain the qf of X (for $0 < u < 1$) as

$$Q(u) = Q_G\left(\exp[-Q^{-1}(\alpha, u)]\right), \tag{7.33}$$

where $z = Q^{-1}(a, u)$ is the inverse function of $Q(a, z) = 1 - \gamma(a, z)/\Gamma(a) = u$. One can also use (7.33) for simulating RB-G variates: if U is a uniform random variable on the unit interval $(0, 1)$, then

$$X = Q_G\left(\exp\left[-Q^{-1}(\alpha, U)\right]\right)$$

will have the distribution defined in equation (7.32).

Next, we provide three special RB-G models.

- *The RB-normal (RBN) distribution*

  The RBN density is obtained from (7.32) by taking $G(\cdot)$ and $g(\cdot)$ to be the cdf and pdf of the normal $N(\mu, \sigma^2)$ distribution. Then, the RBN density function has the form

$$f(x) = \frac{1}{\sigma\,\Gamma(\alpha)}\left\{-\log\left[\Phi\left(\frac{x-\mu}{\sigma}\right)\right]\right\}^{\alpha-1}\phi\left(\frac{x-\mu}{\sigma}\right), \quad x \in \mathbb{R},$$

  where $\mu \in \mathbb{R}$, $\sigma > 0$. For $\mu = 0$ and $\sigma = 1$, we obtain the standard RBN distribution.

- *The RB-logistic (RBLo) distribution*

  We take the baseline logistic distribution with pdf and cdf (for $x \in \mathbb{R}$) given by $g(x) = e^{-x}/(1 + e^{-x})^2$ and $G(x) = (1 + e^{-x})^{-1}$, respectively. The RBLo density function becomes

$$f(x) = \frac{e^{-x}}{\Gamma(\alpha)\,(1 + e^{-x})^2}\left\{\log(1 + e^{-x})\right\}^{\alpha-1}, \quad x \in \mathbb{R}.$$

- *The RB-Weibull (RBW) distribution*

  The baseline Weibull distribution has pdf and cdf (for $x > 0$ and $\lambda, \beta > 0$) $g(x) = \lambda \beta x^{\beta-1} e^{-\lambda x^\beta}$ and $G(x) = 1 - e^{-\lambda x^\beta}$, respectively. Then, the RBW density function reduces to

$$f(x) = \frac{\lambda \beta}{\Gamma(\alpha)} x^{\beta-1} e^{-\lambda x^\beta} \{-\log[1 - e^{-\lambda x^\beta}]\}^{\alpha-1}, \quad x > 0.$$

  The RBW distribution includes as a special case the RB-exponential (RBE) model when $\beta = 1$.

The pdf $f(x)$ in (7.32) can be expressed using (7.5) and similar algebra of Section 7.3 as

$$f(x) = \sum_{k=0}^{\infty} d_k \, h_{k+1}(x), \qquad (7.34)$$

where $h_{k+1}(x)$ is the exp-G pdf with power parameter $k+1$, and the coefficients d_k's (for $k \geq 0$) are given by

$$d_k = \frac{(-1)^k}{(k+1)\,\Gamma(\alpha)} \left[\binom{\alpha-1}{k} + (\alpha-1) \sum_{i=0}^{\infty} q_i(\alpha-1) \binom{\alpha+i}{k} \right].$$

Here, the quantities $q_i(\alpha - 1)$ (for $i \geq 0$) are defined by (7.5). Equation (7.34) reveals that the RB-G density function is a linear combination of exp-G densities. It has the same form of equation (7.6) by changing d_k by b_k and setting $a = 1$.
Thus, general mathematical properties for the gamma-G family also hold for the RB-G family.

7.15. ESTIMATION AND APPLICATION

In this section, we investigate the MLEs of the parameters of the RB-G family from complete samples only. Let $x_1, \ldots, x_n$ be the observed values from this family with parameters α and $\boldsymbol{\xi}$. Let $\boldsymbol{\theta} = (\alpha, \boldsymbol{\xi}^\top)^\top$ be the $p \times 1$ parameter vector. The total log-likelihood function for $\boldsymbol{\theta}$ is

$$\ell(\boldsymbol{\theta}) = \sum_{i=1}^{n} \log[g(x_i; \boldsymbol{\xi})] + (\alpha - 1) \sum_{i=1}^{n} \log\{-\log[G(x_i; \boldsymbol{\xi})]\} - n \, \log[\Gamma(\alpha)].$$

The components of the score function $U(\boldsymbol{\theta}) = (U_\alpha, U_{\boldsymbol{\xi}})^\top$ are

$$U_\alpha = \sum_{i=1}^n \log\{-\log[G(x_i; \boldsymbol{\xi})]\} - n\,\psi(\alpha)$$

and

$$U_{\boldsymbol{\xi}} = \sum_{i=1}^n \frac{\partial g(x_i; \boldsymbol{\xi})/\partial\boldsymbol{\xi}}{g(x_i; \boldsymbol{\xi})} + (\alpha - 1) \sum_{i=1}^n \frac{\partial G(x_i; \boldsymbol{\xi})/\partial\boldsymbol{\xi}}{G(x_i; \boldsymbol{\xi})\,\log[G(x_i; \boldsymbol{\xi})]},$$

where $\psi(\alpha) = d\log[\Gamma(\alpha)]/d\alpha$ is the digamma function.

We can obtain the MLEs $\widehat{\boldsymbol{\theta}} = (\widehat{\alpha}, \widehat{\boldsymbol{\xi}})^\top$ by setting U_α and $U_{\boldsymbol{\xi}}$ equal to zero. These equations cannot be solved analytically but the estimates can be evaluated numerically using Newton-Raphson type algorithms.

The MLE of α, denoted by $\widehat{\alpha}$, is the solution of the following non-linear equation

$$\psi(\widehat{\alpha}) = \frac{1}{n} \sum_{i=1}^n \log[\log(1 + e^{-x_i})].$$

For interval estimation on the model parameters, we require the observed information matrix

$$J(\boldsymbol{\theta}) = -\left(\begin{array}{c|c} U_{\alpha\alpha} & U_{\alpha\boldsymbol{\xi}}^\top \\ \hline U_{\alpha\boldsymbol{\xi}} & U_{\boldsymbol{\xi}\boldsymbol{\xi}} \end{array}\right),$$

whose elements are

$$U_{\alpha\alpha} = -n\,\psi'(\alpha), \quad U_{\alpha\boldsymbol{\xi}_k} = \sum_{i=1}^n \frac{\partial G(x_i; \boldsymbol{\xi})/\partial\boldsymbol{\zeta}}{G(x_i; \boldsymbol{\xi})\,\log[G(x_i; \boldsymbol{\xi})]},$$

and

$$U_{\boldsymbol{\xi}_k\boldsymbol{\xi}_l} = (\alpha - 1)\sum_{i=1}^n \{G(x_i; \boldsymbol{\xi})\,\log[G(x_i; \boldsymbol{\xi})]\}^{-2}\{G_{kl}''(x_i)\,G(x_i; \boldsymbol{\xi})\,\log[G(x_i; \boldsymbol{\xi})]$$

$$- G_k'(x_i)\,G_l'(x_i)[\log[G(x_i)] + 1]\} + \sum_{i=1}^n \frac{g_{kl}''(x_i)g(x_i) - g_k'(x_i)g_l'(x_i)}{[g(x_i)]^2},$$

where $t_k'(\cdot; \boldsymbol{\xi}) = \partial t(\cdot; \boldsymbol{\xi})/\partial\boldsymbol{\xi}_k$ and $t_{kl}''(\cdot; \boldsymbol{\xi}) = \partial^2 t(\cdot; \boldsymbol{\xi})/\partial\boldsymbol{\xi}_k\partial\boldsymbol{\xi}_l^\top$.

Approximate confidence intervals for the parameters can be based on the multivariate normal $N_{p+1}(0, J(\widehat{\boldsymbol{\theta}}))$ distribution, where $J(\widehat{\boldsymbol{\theta}})$ is the observed information matrix evaluated at $\widehat{\boldsymbol{\theta}}$. Its elements can be obtained from the authors upon request. We can compute the maximum values of the unrestricted and restricted log-likelihoods to obtain LR statistic for testing the RB-G model against the G model.

Next, we provide an application to the failure times of 20 mechanical components reported in Murthy *et al.* (2004) [57]. The data are:

0.067, 0.068, 0.076, 0.081, 0.084, 0.085, 0.085, 0.086, 0.089, 0.098, 0.098, 0.114, 0.114, 0.115, 0.121, 0.125, 0.131, 0.149, 0.160, 0.485.

We consider the RB-log-logistic (RBLL) distribution obtained from (7.32) as

$$f(x) = \frac{c}{s^c\,\Gamma(\alpha)}\,x^{c-1}\left[1+(x/s)^c\right]^{-2}\left\{-\log(1-[1+(x/s)^c]^{-1})\right\}^{\alpha-1}$$

and compare its fit to these data with the beta log-logistic (BLL) (Lemonte, 2014 [125]), Kumaraswamy log-logistic (KwLL) (Santana *et al.*, 2012 [191]), Zografos-Balakrishnan log-logistic (ZBLL) (Ramos *et al.*, 2013 [206]), exponentiated log-logistic (ELL) and log-logistic (LL) models with corresponding densities (for $x > 0$)

$$\text{BLL}: f_{\text{BLL}}(x; s, c, \alpha, \beta) = \frac{(c/s)}{B(\alpha, \beta)}\frac{(x/s)^{\alpha c-1}}{[1+(x/s)]^{\alpha+\beta}},$$

$$\text{KwLL}: f_{\text{KwLL}}(x; s, c, \alpha, \beta) = \frac{\alpha\,\beta\,c\,(x/s)^{c\alpha-1}}{s[1+(x/s)^c]^{\alpha+1}}\left\{1-\frac{(x/s)^{c\alpha}}{[1+(x/s)^c]^\alpha}\right\}^{\beta-1},$$

$$\text{ZBLL}: f_{\text{ZBLL}}(x; s, c, \alpha) = \frac{c\,(x/s)^{c-1}}{s\,\Gamma(\alpha)}\frac{\{\log[1+(x/s)^c]\}^{\alpha-1}}{[1+(x/s)^c]^2},$$

$$\text{ELL}: f_{\text{ELL}}(x; s, c, \alpha) = \frac{\alpha\,c\,(x/s)^{c\alpha-1}}{s[1+(x/s)^c]^{\alpha+1}},$$

where $s > 0, c > 0, \alpha > 0$ and $\beta > 0$.

We estimate the parameters of these models by the maximum likelihood method. All computations are carried out using the **AdequacyModel** script of the R-package freely available from `http://cran.r-project.org/web/packages/AdequacyModel/AdequacyModel.pdf`.

Table 7.2 lists the MLEs (with corresponding SEs in parentheses) and the AIC, BIC and CAIC statistics. Since the values of these statistics are smaller for the RBLL and ZBLL distributions compared with those values of the other models, these distributions are very competitive to the current data.

Table 7.2: MLEs of the model parameters, the corresponding SEs (given in parentheses) and the AIC, BIC and CAIC statistics.

Model	Estimates				Statistic		
	s	c	α	β	AIC	BIC	CAIC
BLL	0.0682	4.3499	2.8026	0.6797	−67.638	−63.655	−64.971
	(0.0157)	(1.2501)	(2.1128)	(0.2895)			
KwLL	0.0651	3.3872	3.5040	0.9992	−66.563	−62.580	−63.896
	(0.0137)	(1.1846)	(1.9384)	(0.5451)			
ELL	0.0718	4.1517	3.1796		−69.744	−66.757	−68.244
	(0.0160)	(0.8341)	(2.1703)				
RBLL	0.0812	6.7272	0.4741		**−70.279**	**−67.292**	**−68.779**
	(0.0109)	(2.3075)	(0.2541)				
ZBLL	0.0603	4.9632	2.9333		−69.368	−66.381	−67.868
	(0.0133)	(1.0760)	(1.1769)				
LL	0.1012	5.0655			−67.649	−66.363	−67.649
	(0.0075)	(0.9591)					

Table 7.3 provides the values of the Cramér-von Mises (CM) and Anderson-Darling (AD) statistics for these data. Thus, according to these formal tests, the RBLL model fits the studied data better than the other models. Curves of the estimated pdfs and cdfs due to two best fitted models are exhibited in Figure 7.2.

Table 7.3: Goodness-of-fit tests.

Model	Statistics	
	CM	AD
BLL	0.0538	0.4110
KLL	0.0638	0.4927
ELL	0.0575	0.4407
RBLL	**0.0515**	**0.3849**
ZBLL	0.0610	0.4758
LL	0.0877	0.6588

7.16.   CONCLUSIONS

We study general properties of two classes of distributions called the *gamma*-G (Zografos and Balakrishnan, 2009 [201]) and *Ristić-Balakrishnan*-G ("RB-G")

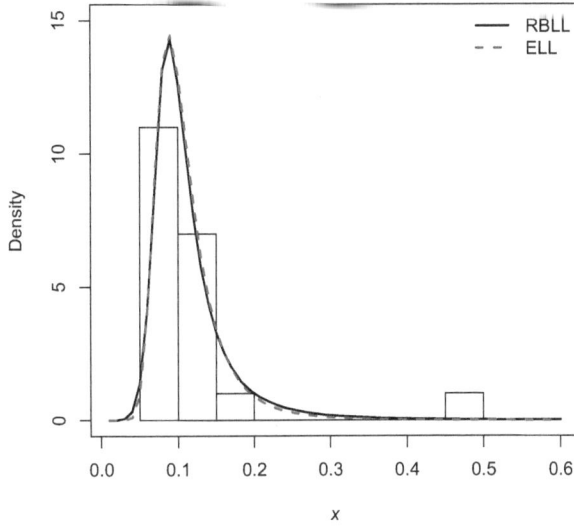

(a) Empirical and theoretic pdfs

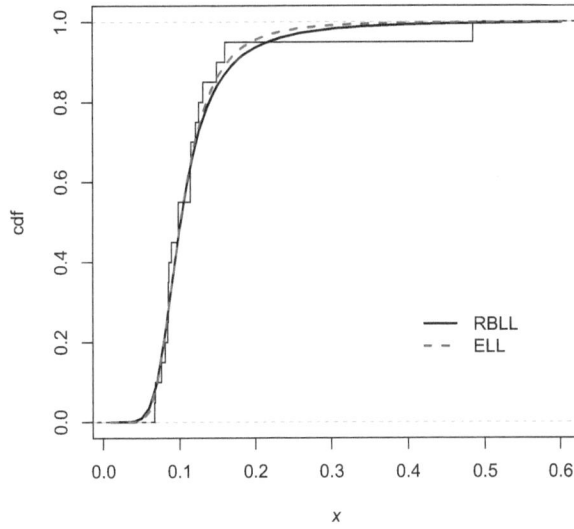

(b) Empirical and theoretic cdfs

Figure 7.2: Estimated (a) pdfs and (b) cdfs for the two best fitted models to the failure time data.

families, which include as special cases all classical continuous distributions. For any parent continuous distribution G, we define the corresponding gamma-G and ZB-G distributions with an extra positive parameter. So, these classes can extend   several well-known distributions such as the normal, Weibull, log-normal, Gumbel and log-logistic distributions. Some structural properties of these classes such as the ordinary and incomplete moments, generating and quantile functions, mean deviations, Bonferroni and Lorenz curves, Shannon entropy, Rényi entropy and order statistics are investigated. The model parameters are estimated by maximum likelihood. Examples to real data illustrate the potentiality of the two classes.

Chapter 8

Recent Compounding Models

Abstract: In this chapter, we introduce a family of   models defined by compounding two (a continuous and other discrete) distributions. The new family has as limiting case the adopted baseline distribution. The generated models are frequently more flexible than the baseline distributions. Several mathematical properties such as moments, quantile and generating functions, among others, are provided. Further, the estimation procedure is approched by maximum likelihood. The potentiality of the family of models is illustrated by means of two applications to real data.

Keywords: BSPS; BXIIPS; Compounding Models; EWPS; Generating function; Moment; Order statistic; WPS.

8.1.   INTRODUCTION

In this chapter, we review some recent compounding lifetime distributions, which were pioneered by Marshall and Olkin (1998) [4] and after extended by some authors. Several well-known lifetime models, such as the exponential, gamma and Weibull distributions, have been extended by compounding lifetime distributions recently introduced in the statistical literature. The class of compounding distributions allows for the use in industrial applications and biological research. It arises by mixing the power series and lifetime distributions. It is specially useful in a situation "where the failure occurs due to the presence of an unknown number, say N, of initial defects of the same kind and the T's represent their lifetimes and each defect can be detected only after causing failure, in which case it is repaired perfectly" (Adamidis and Loukas, 1998 [208]).

Gauss M. Cordeiro, Rodrigo B. Silva & Abraão D. C. Nascimento

Given N, let $T_1, \ldots, T_N$ be iid random variables having a baseline cdf $G(t; \boldsymbol{\tau})$, where $\boldsymbol{\tau}$ is a vector of parameters and N is a discrete random variable following a zero truncated power series (PS) distribution with probability mass function (pmf) expressed by

$$p_n = P(N = n) = \frac{a_n \, \theta^n}{C(\theta)}, \quad n = 1, 2, \ldots. \tag{8.1}$$

Note that the coefficient a_n is a function of n and $C(\theta) = \sum_{n=1}^{\infty} a_n \, \theta^n$, with $\theta > 0$ such that $C(\theta)$ is finite. It is important to emphasize that the probability distributions of the form (8.1) have been considered in Boehme and Powell (1968) [208] and Ostrovska (2007) [209]. Table 8.1 lists some PS distributions defined by (8.1) such as the Poisson, logarithmic, geometric and binomial distributions.

Define $X = \min \{T_i\}_{i=1}^{N}$. The conditional cumulative distribution of $X|N = n$ is given by

$$F_{X|N=n}(x) = 1 - [1 - G(x; \boldsymbol{\tau})]^n$$

and then

$$P(X \leq x, N = n) = \frac{a_n \theta^n}{C(\theta)} \left\{ 1 - [1 - G(x; \boldsymbol{\tau})]^n \right\}, \quad x > 0, \quad n = 1, 2, \ldots.$$

Therefore, the marginal cdf of X becomes

$$F(x; \theta, \boldsymbol{\tau}) = 1 - \frac{1}{C(\theta)} C \left\{ \theta \left[1 - G(x; \boldsymbol{\tau}) \right] \right\}, \quad x > 0. \tag{8.2}$$

Distribution	a_n	$C(\theta)$	$C'(\theta)$	$C''(\theta)$	$C(\theta)^{-1}$	Θ
Poisson	$n!^{-1}$	$e^{\theta} - 1$	e^{θ}	e^{θ}	$\log(\theta + 1)$	$\theta \in (0, \infty)$
Logarithmic	n^{-1}	$-\log(1 - \theta)$	$(1 - \theta)^{-1}$	$(1 - \theta)^{-2}$	$1 - e^{-\theta}$	$\theta \in (0, 1)$
Geometric	1	$\theta(1 - \theta)^{-1}$	$(1 - \theta)^{-2}$	$2(1 - \theta)^{-3}$	$\theta(\theta + 1)^{-1}$	$\theta \in (0, 1)$
Binomial	$\binom{m}{n}$	$(\theta + 1)^m - 1$	$m(\theta + 1)^{m-1}$	$\frac{m(m-1)}{(\theta+1)^{2-m}}$	$(\theta - 1)^{1/m} - 1$	$\theta \in (0, 1)$

Table 8.1: Functional quantities for some PS distributions.

The pdf associated to (8.2) is

$$f(x; \theta, \boldsymbol{\tau}) = \frac{\theta}{C(\theta)} g(x; \boldsymbol{\tau}) C' \left\{ \theta \left[1 - G(x; \boldsymbol{\tau}) \right] \right\}, \quad x > 0, \tag{8.3}$$

where $g(x;\boldsymbol{\tau})$ is the baseline density function and $C'(\cdot)$ is the first derivative with respect to θ.

The random variable X with density (8.3) is called the *G-power series* (GPS) family and denoted by $X \sim GPS(\theta,\boldsymbol{\tau})$, which is customary for such a name given to the distributions arising by means of the operation of compounding.

Remark. In an analogous way for X, define $Y = \max\{T_i\}_{i=1}^{N}$, where $N \sim PS(\theta)$. Then, the cumulative and density functions of Y are

$$F(y;\theta,\boldsymbol{\tau}) = \frac{1}{C(\theta)} C\left[\theta\,G(y;\boldsymbol{\tau})\right], \quad y > 0$$

and

$$f(y;\theta,\boldsymbol{\tau}) = \frac{\theta}{C(\theta)} g(y;\boldsymbol{\tau})\,C'\left[\theta\,G(y;\boldsymbol{\tau})\right], \tag{8.4}$$

respectively. The family with cumulative distribution (8.4) is a complement of the GPS family and thus, hereafter, it is called the *complementary G-power series* (CGPS) family, denoted by $Y \sim \text{CGPS}(\theta,\boldsymbol{\tau})$.

This type of compounding family is suitable for complementary risks scenarios, where the lifetime corresponding to a particular risk is not perceptible, rather we observe only the maximum lifetime value among all risks. Note that equations (8.3) and (8.4) will be most manageable when both functions $G(x;\boldsymbol{\tau})$ and $g(x;\boldsymbol{\tau})$ have uncomplicated expressions. In general, except for some special choices of these functions, these densities will be difficult to deal with. A positive point of the compounding distributions is that the baseline distribution G is a basic exemplar of the generated model. In addition, the compounding distributions have various interesting applications based on the stochastic representations (8.3) and (8.4), which make them of recognizable scientific relevance from other lifetime distributions. We list below some of these interesting applications.

- *Time to the first failure* (Adamidis and Loukas, 1998 [208] and Kus, 2007 [211]). Consider that a component or system can fail after the ocurrence of a number N of early defects of the same kind, only detected after causing failure and perfectly repaired. If we denote by T_i the time to the device failure due to the ith defect, then the model defined in (8.3) is adequate for modelling the time to the first failure, under the assumptions that the T_i's are iid random variables independent of N, which is defined in (8.1).

- *Reliability.* From the stochastic quantities (8.3) and (8.4), we have that the compounding models can emerge in series and parallel systems with identical components, which appear in many industrial applications and biological organisms.

- *Time to relapse of cancer under the first-activation scheme* (Chen *et al.*, 1999 [211]). In this case, suppose that an individual is susceptible to a certain type of cancer. Consider that the number, N, of carcinogenic cells for that individual still active after a initial treatment has a PS distribution and let T_i be the time spent for the ith carcinogenic cell to produce a detectable cancer mass, for $i \geq 1$. Then, the time to relapse of cancer of a susceptible individual can be modelled by a distribution defined by (8.3).

- *Last-activation scheme* (Cooner *et al.*, 2007 [212]). Assume that the number, say N, of potential factors that must be activated by failure has a PS distribution. Further, suppose that T_i denotes the resistance time to a disease (defect) manifestation due to the ith potential factor following a lifetime distribution. Under the last-activation scheme, the failure occurs after all N factors be activated. So, the stochastic representation (8.4) is suitable for modelling the time to the failure in this scheme.

The compounding distributions have received considerable attention over the last years, in particular after the work of Adamidis and Loukas (1998) [207], where they pionnered the exponential geometric (EG) distribution. Since the publication of this seminal paper, many other generalizations have been proposed. Some of them are described below.

The EG distribution is given by compounding the exponential and zero truncated geometric distributions. Suppose that $\{T_i\}_{i=1}^{N}$ are iid random variables with density $f(t_i; \lambda) = \lambda e^{-\lambda t_i}$ and N is a zero truncated geometric variable with parameter $\theta \in (0, 1)$, see Table 8.1. Let $X = \min \{T_i\}_{i=1}^{N}$. Then,

$$f(x; \theta, \lambda) = \lambda (1 - \theta) e^{-\lambda t} (1 - \theta e^{-\lambda t})^{-2}, \quad x > 0.$$

The above density defines the distribution that is referred to as the EG distribution. Several properties of the EG model were discussed and the estimation procedure was performed by an EM algorithm, in addition to the maximum likelihood approach.

In a similar manner, the exponential Poisson (EP) and exponential logarithmic (EL) distributions were introduced and studied by Kuş (2007) [207] and Tahmasbi and Rezaei (2008) [213], respectively. These works were generalized by Chahkandi and Ganjali (2009) [35], who showed that the composition of the exponential and PS distributions yields a distribution with a decreasing failure rate, the exponential power series (EPS) distribution. This is another seminal paper, since its methodology was used for introducing several other families of compounding lifetime distributions. This family is defined by combining the exponential and PS distributions in Table 8.1. So, let $X = \min \{T_i\}_{i=1}^{N}$, where $T_1, \ldots, X_N$ are iid exponential random variables, and let N follow any

PS distribution in Table 8.1. Then, $P(X \le x | N = n) = 1 - \mathrm{e}^{-n\lambda x}$ and the marginal pdf of X has the form

$$f(x; \theta, \lambda) = \frac{\lambda \theta}{C(\theta)} \mathrm{e}^{-\lambda x} C'(\theta \, \mathrm{e}^{-\lambda x}), \quad x > 0,$$

for $0 < \theta < 1$ and $\lambda > 0$. The above density defines the family of distributions referred to as the EPS distributions. Chahkandi and Ganjali (2009) [35] provided expressions for the moments and examined the estimation procedure by maximum likelihood and by an EM algorithm. Further extensions of the EP and EG distributions were proposed by Barreto-Souza and Cribari-Neto (2009) [33] and Silva *et al.* (2010) [9]. Since the Weibull distribution extends the exponential distribution, it is straightforward to extend the EG distribution by replacing the exponential by the Weibull distribution in the compounding mechanism. Following this point, Barreto-Souza *et al.* (2010) [168] proposed the Weibull geometric (WG) distribution and studied various of its properties.

In an analogous way, Lu and Shi (2012) [215] proposed the Weibull Poisson (WP) distribution, which naturally extends the EP distribution. Following the same idea of Chahkandi and Ganjali (2009) [35], Morais and Barreto-Souza (2011) [216] introduced the Weibull power series (WPS) family, containing the EPS distributions as particular models.

In Section 8.3, we prove that the general WPS density function can be expressed as a linear combination of Weibull densities. Hence, some mathematical properties of the WPS models can be obtained from those of the Weibull distribution. Following the same approach developed by Chahkandi and Ganjali (2009) [35] and Morais and Barreto-Souza (2011) [216], Mahmoudi and Jafari (2012) introduced generalized exponential power series (GEPS) distributions. Recent compounded distributions were investigated by Cancho *et al.* (2011 [217], 2012 [218]), who proposed the Poisson exponential (PE) and geometric Birnbaum-Saunders (GBS) distributions and Barreto-Souza and Bakouch (2013) [219], who introduced the Poisson Lindley (PL) distribution. Silva *et al.* (2013) [220] defined the extended Weibull power series (EWPS) family, containing the EPS and WPS distributions as special cases. The EWPS distributions are obtained by mixing the extended Weibull (EW) models and PS distributions. The EW family was defined by Gurvich *et al.* (1997) [221] in the context of modelling failure times and random strength of brittle materials. The EWPS cdf can be expressed as

$$G(t; \alpha, \boldsymbol{\tau}) = 1 - \exp\left\{-\alpha \, H(t; \boldsymbol{\tau})\right\}, \quad x > 0, \ \alpha > 0, \tag{8.5}$$

where $\boldsymbol{\tau}$ is a vector of parameters and $H(t; \boldsymbol{\tau})$ is a non-negative monotonically increasing function. The associated density function becomes

$$g(t; \alpha, \boldsymbol{\tau}) = \alpha \, h(t; \boldsymbol{\tau}) \exp \left\{ -\alpha \, H(t; \boldsymbol{\tau}) \right\}, \quad x > 0, \ \alpha > 0, \qquad (8.6)$$

where $h(t; \boldsymbol{\tau})$ is the first derivative of $H(t; \boldsymbol{\tau})$. Many known models can be expressed in the form (8.5) such as those in Table 8.2, where some functional quantities and corresponding parameter vectors are given. Hence, the EWPS family defines 68 special cases, which can have increasing, decreasing, bathtub and upside down bathtub hrfs. They demonstrated that the EW family with parameters α and $\boldsymbol{\tau}$ is a limiting special case of the EWPS family when $\theta \to 0^+$. Further, they proved that the EWPS densities can be linear representations of EW densities, which allow for the EWPS models to construct some of their properties based on those of the EW family. Bourguignon *et al.* (2013) [221] and Silva and Cordeiro (2015) [222] introduced and studied the Birnbaum-Saunders power series (BSPS) and Burr XII power series (BXIIPS) distributions, respectively.

Based on equation (8.4), a distinct method, which considers the maximum as an alternative to the minimum, has also been considered. Louzada *et al.* (2011) [230] introduced the complementary exponential geometric (CEG) distribution, which is the counterpart of the EG distribution defined by Adamidis and Loukas (1998) [207]. In terms of illustration, they fitted the CEG distribution to lifetimes of 23 ball bearings on an endurance test of deep-groove ball bearings. Cordeiro *et al.* (2012) [231] proposed the exponential COM Poisson (ECOMP) distributions, which extends the Poisson-exponential (PE) lifetime distribution by Cancho *et al.* (2011) [216]. The complementary exponential power series (CEPS) distributions were defined and many of their properties studied by Flores *et al.* (2013) [232]. The CEPS models are the counterpart of the EPS distributions and their properties include increasing hrfs and the density function as an infinite mixture of densities of the maximum order statistics of the baseline distribution. In a very recent paper, Cordeiro and Silva (2015) [222] introduced and studied several properties of the complementary extended Weibull power series (CEWPS) family. In a recent paper, Barreto-Souza and Silva (2015) [233] proposed a likelihood ratio test based on Cox's statistic to discriminate between the EP and gamma distributions. The asymptotic distribution of the normalized logarithm of the ratio of the maximized likelihoods under two null hypotheses - data come from EP or gamma distributions - was provided. For doing this, they obtained the probabilities of correct selection.

Some special compounding models are given in Table 8.3. For illustrative purposes, we now provide the cumulative and density functions of four GPS distributions:

Table 8.2: Extended Weibull distributions and corresponding $H(x;\tau)$ and $h(x;\tau)$ functions.

Distribution	$H(x;\tau)$	$h(x;\tau)$	α	τ	References
Exponential ($x \geq 0$)	x	1	α	$\emptyset$	Johnson et al. (1994) [223]
Pareto ($x \geq k$)	$\log(x/k)$	$1/x$	α	k	Johnson et al. (1994) [223]
Rayleigh ($x \geq 0$)	x^2	$2x$	α	$\emptyset$	Rayleigh (1880) [224]
Weibull ($x \geq 0$)	x^γ	$\gamma x^{\gamma-1}$	α	γ	Johnson et al. (1994) [223]
Modified Weibull ($x \geq 0$)	$x^\gamma \exp(\lambda x)$	$x^{\gamma-1} \exp(\lambda x)(\gamma + \lambda x)$	α	(γ, λ)	Lai et al. (2003) [72]
Weibull extension ($x \geq 0$)	$\lambda[\exp(x/\lambda)^\beta - 1]$	$\beta \exp(x/\lambda)^\beta (x/\lambda)^{\beta-1}$	α	(γ, λ, β)	Xie et al. (2002) [156]
Log-Weibull ($-\infty < x < \infty$)	$\exp[(x-\mu)/\sigma]$	$(1/\sigma)\exp[(x-\mu)/\sigma]$	1	(μ, σ)	White (1969) [225]
Phani ($0 < \mu < x < \sigma < \infty$)	$[(x-\mu)/(\sigma-x)]^\beta$	$\beta[(x-\mu)/(\sigma-x)]^{\beta-1}[(\sigma-\mu)/(\sigma-t)^2]$	α	(μ, σ, β)	Phani (1987) [226]
Weibull Kies ($0 < \mu < x < \sigma < \infty$)	$(x-\mu)^{\beta_1}/(\sigma-x)^{\beta_2}$	$(x-\mu)^{\beta_1-1}(\sigma-x)^{-\beta_2-1}[\beta_1(\sigma-x) + \beta_2(x-\mu)]$	α	$(\mu, \sigma, \beta_1, \beta_2)$	Kies (1958) [227]
Additive Weibull ($x \geq 0$)	$(x/\beta_1)^{\alpha_1} + (x/\beta_2)^{\alpha_2}$	$(\alpha_1/\beta_1)(x/\beta_1)^{\alpha_1-1} + (\alpha_2/\beta_2)(x/\beta_2)^{\alpha_2-1}$	1	$(\alpha_1, \alpha_2, \beta_1, \beta_2)$	Xie and Lai (1995) [171]
Traditional Weibull ($x \geq 0$)	$x^b[\exp(cx^d) - 1]$	$bx^{b-1}[\exp(cx^d) - 1] + cdx^{b+d-1}\exp(cx^d)$	α	(b, c, d)	Nadarajah and Kotz (2005) [41]
Gen. power Weibull ($x \geq 0$)	$[1 + (x/\beta)^{\alpha_1}]^\theta - 1$	$(\theta\alpha_1/\beta)[1 + (x/\beta)^{\alpha_1}]^{\theta-1}(x/\beta)^{\alpha_1}$	1	$(\alpha_1, \beta, \theta)$	Nikulin and Haghighi (2006) [75]
Flexible Weibull extension ($x \geq 0$)	$\exp(\alpha_1 x - \beta/x)$	$\exp(\alpha_1 x - \beta/x)(\alpha_1 + \beta/x^2)$	1	(α_1, β)	Bebbington et al. (2007) [166]
Gompertz ($x \geq 0$)	$\beta^{-1}[\exp(\beta x) - 1]$	$\exp(\beta x)$	α	β	Gompertz (1825) [55]
Exponential power ($x \geq 0$)	$\exp[(\lambda x)^\beta] - 1$	$\beta\lambda \exp[(\lambda x)^\beta](\lambda x)^{\beta-1}$	1	(λ, β)	Smith and Bain (1975) [228]
Chen ($x \geq 0$)	$\exp(x^b) - 1$	$bx^{b-1}\exp(x^b)$	α	b	Chen (2000) [165]
Pham ($x \geq 0$)	$(a^x)^\beta - 1$	$\beta(a^x)^\beta \log(a)$	1	(a, β)	Pham (2002) [229]

- The cumulative and density functions of the EP distribution (Kuş, 2007 [210]) are:

$$F(x;\theta,\lambda) = 1 - \frac{\exp(\theta\,e^{-\lambda x}) - 1}{e^\theta - 1}, \quad x > 0$$

and

$$f(x;\theta,\lambda) = \frac{\lambda\,\theta}{e^\theta - 1}\exp\left\{-\lambda x + \theta\,e^{-\lambda x}\right\}, \quad x > 0,$$

where $\lambda > 0$ is a scale parameter and $0 < \theta < 1$ is a shape parameter.

- The cumulative and density functions of the WG distribution (Barreto-Souza *et al.*, 2010 [234]) are:

$$F(x;\theta,\alpha,\beta) = \frac{1 - e^{-(\beta x)^\alpha}}{1 - \theta\,e^{-(\beta x)^\alpha}}, \quad x > 0$$

and

$$f(x;\theta,\alpha,\beta) = \alpha\beta^\alpha(1-\theta)x^{\alpha-1}e^{-(\beta x)^\alpha}\left\{1 - \theta\,e^{-(\beta x)^\alpha}\right\}^{-2}, \quad x > 0,$$

where $\beta > 0$ is a scale parameter and $0 < \theta < 1$ and $\alpha > 0$ are shape parameters.

- The cumulative and density functions of the *Chen logarithmic* distribution (Silva *et al.*, 2013 [219]) are:

$$F(x;\theta,\alpha,\beta) = 1 - \frac{\log\left\{1 - \theta\exp\left[-\alpha(e^{x^b} - 1)\right]\right\}}{\log(1 - \theta)}, \quad x > 0$$

and

$$f(x;\theta,\alpha,\beta) = \frac{\theta\,\alpha\,b\,x^{b-1}\exp\left\{x^b - \alpha\left[e^{x^b} - 1\right]\right\}}{\log(1 - \theta)\left\{\theta\exp\left[-\alpha(e^{x^b} - 1)\right] - 1\right\}}, \quad x > 0,$$

where $\alpha > 0$ and $0 < \theta < 1$ are scale parameters and $b > 0$ is a shape parameter.

- The cumulative and density functions of the CEG distribution (Louzada *et al.*, 2011 [230]) are:

$$F(x;\theta,\lambda) = 1 - \frac{e^{-\lambda x}}{e^{-\lambda x}(1 - \theta) + \theta}, \quad x > 0$$

and

$$f(x;\theta,\lambda) = \frac{\lambda\,\theta\,e^{-\lambda x}}{[e^{-\lambda x}(1 - \theta) + \theta]^2}, \quad x > 0,$$

where $\lambda > 0$ and $0 < \theta < 1$ are scale parameters.

Table 8.3: Some special GPS distributions. The function $g(\cdot)$ is the parent density function.

Distribution	Baseline Distribution	GPS density
Exponential	$G(x;\lambda)=1-\mathrm{e}^{-\lambda x}$	$f(x;\theta,\lambda)=\dfrac{\lambda\theta}{C(\theta)}\,\mathrm{e}^{-\lambda x}C'(\theta\mathrm{e}^{-\lambda x})$
Weibull	$G(x;\theta,\alpha,\beta)=1-\mathrm{e}^{-(\beta x)^\alpha}$	$f(x;\theta,\alpha,\beta)=\dfrac{\theta\alpha\beta^\alpha}{C(\theta)}\,x^{\alpha-1}\mathrm{e}^{-(\beta x)^\alpha}C'(\theta\mathrm{e}^{-(\beta x)^\alpha})$
Burr XII	$G(x;c,k)=1-(1+x^c)^{-k}$	$f(x;c,k)=\dfrac{\theta ck}{C(\theta)}\,x^{c-1}(1+x^c)^{-(k+1)}C'[\theta(1+x^c)^{-k}]$
Birnbaum-Saunders	$G(x;\alpha,\beta)=\Phi\left[\dfrac{1}{\alpha}\left(\sqrt{\dfrac{x}{\beta}}-\sqrt{\dfrac{\beta}{x}}\right)\right]$	$f(x;\theta,\alpha,\beta)=\dfrac{\theta}{C(\theta)}g(x;\alpha,\beta)C'\{\theta[1-\Phi(v)]\}$
Modified Weibull	$G(x;\alpha,\beta,\gamma)=1-\mathrm{e}^{-\alpha x-\beta x^\gamma}$	$f(x;\theta,\alpha,\beta,\gamma)=\dfrac{\theta}{C(\theta)}g(x;\alpha,\beta,\gamma)C'\{\theta\mathrm{e}^{-\alpha x-\beta x^\gamma}\}$
Generalized gamma	$G(x;\alpha,\beta,k)=\dfrac{\gamma(k,(x/\beta)^\alpha)}{\Gamma(k)}$	$f(x;\theta,\alpha,\beta,k)=\dfrac{\theta}{C(\theta)}g(x;\alpha,\beta,k)C'\left\{\theta\left[1-\gamma_1\left(k,\left(\dfrac{x}{\beta}\right)^\alpha\right)\right]\right\}$

8.2. QUANTILE FUNCTION

The qf of the GPS family, say $Q(u) = Q(u; \theta, \boldsymbol{\tau})$, can be obtained by inverting the parent G cumulative distribution. Then, it can be expressed in terms of the baseline qf as

$$Q(u) = G^{-1}\left\{1 - \frac{C^{-1}[(1-u)C(\theta)]}{\theta}\right\}, \quad u \in (0,1), \tag{8.7}$$

where $G^{-1}(\cdot)$ and $C^{-1}(\cdot)$ are the inverse functions of $G(\cdot)$ and $C(\cdot)$, respectively. In the same way, we have

$$Q(u) = G^{-1}\left\{\frac{C^{-1}[u\,C(\theta)]}{\theta}\right\}, \quad u \in (0,1), \tag{8.8}$$

for the CGPS family of distributions.

We generate GPS and CGPS random variables following the procedure:

- *Step 1*: Generate $U \sim U(0,1)$;

- *Step 2*: Set values for θ and $\boldsymbol{\tau}$ of $X \sim \text{GPS}(\theta, \boldsymbol{\tau})$ or $X \sim \text{CGPS}(\theta, \boldsymbol{\tau})$;

- *Step 3*: Determine the inverse function $G^{-1}(\cdot)$ according to the parent G cumulative function;

- *Step 4*: Define a function $C^{-1}(\cdot)$ such as described in Table 8.1 and use (8.7) or (8.8) to generate a GPS or CGPS distribution, respectively;

- *Step 5*: Obtain an outcome of X by $X = Q(U)$;

- *Step 6*: Repeat Steps 1 to 5 until the required amount of random numbers be achieved.

To simulate data from the nonlinear equations (8.7) and (8.8), we can use the matrix programming language `Ox` through `SolveNLE` subroutine (see Doornik, 2007 [150]).

8.3. USEFUL EXPANSIONS

We present useful expansions for some compounding density functions. The GPS density function can be written as a mixture of densities of minimum order statistics of the baseline random variable T. Indeed, note that $C'(\theta) = \sum_{n=1}^{\infty} n\, a_n\, \theta^{n-1}$. Then,

$$f(x; \theta, \boldsymbol{\tau}) = \frac{\theta}{C(\theta)} g(x; \boldsymbol{\tau}) \sum_{n=1}^{\infty} n\, a_n\, \theta^{n-1} [1 - G(x; \boldsymbol{\tau})]^{n-1}$$

$$= \sum_{n=1}^{\infty} p_n\, g_{T_{(1)}}(x; \boldsymbol{\tau}), \quad x > 0, \tag{8.9}$$

where $g_{T_{(1)}}(\cdot)$ is the density function of $T_{(1)} = \min\{T_i\}_{i=1}^{n}$, for fixed n, given by

$$g_{T_{(1)}}(t; \boldsymbol{\tau}) = n\, g(t; \boldsymbol{\tau})[1 - G(t; \boldsymbol{\tau})]^{n-1}, \quad t > 0.$$

By using the binomial theorem, we can write equation (8.9) in terms of a linear combination of the exponentiated baseline densities as follows (for $x > 0$)

$$g_{T_{(1)}}(t; \boldsymbol{\tau}) = n\, g(t; \boldsymbol{\tau}) \sum_{j=0}^{n-1} (-1)^j \binom{n-1}{j} G(t; \boldsymbol{\tau})^j$$

$$= n\, g(t; \boldsymbol{\tau}) \sum_{k=1}^{n} (-1)^{k-1} \binom{n-1}{k-1} G(t; \boldsymbol{\tau})^{k-1}$$

$$= \sum_{k=1}^{n} (-1)^{k-1} \binom{n}{k} g_k(t; \boldsymbol{\tau}), \tag{8.10}$$

where the integer k is a power parameter for which the baseline distribution G is raised to originate the exponentiated baseline distribution, whose cumulative function is $G_k(\cdot) = G(\cdot)^k$. Inserting (8.10) in equation (8.9), we have

$$f(x; \theta, \boldsymbol{\tau}) = \sum_{n=1}^{\infty} \sum_{k=1}^{n} (-1)^{k-1} p_n \binom{n}{k} g_k(x; \boldsymbol{\tau}), \quad x > 0.$$

By changing $\sum_{n=1}^{\infty} \sum_{k=1}^{n}$ to $\sum_{k=1}^{\infty} \sum_{n=k}^{\infty}$, we can write

$$f(x; \theta, \boldsymbol{\tau}) = \sum_{k=1}^{\infty} \nu_k\, g_k(x; \boldsymbol{\tau}), \quad x > 0, \tag{8.11}$$

where $\nu_k = (-1)^{k-1} \sum_{n=k}^{\infty} p_n \binom{n}{k}$. There are many established exponentiated distributions in the statistical literature such as the exponentiated exponential (EE) (Gupta and Kundu, 1999 [42]) and exponentiated Weibull (EW) (Mudholkar and Srivastava, 1993 [7]) distributions. So, one can choose between equations (8.9) and (8.11) depending on if it will be difficult to deal with one of them. Table 8.4 lists closed-form expressions of $\sum_{n=k}^{\infty} p_n \binom{n}{k}$ for the Poisson, logarithmic, geometric and binomial distributions. These expressions can be obtained using programs for

Table 8.4: Closed-form expressions for $\sum_{n=k}^{\infty} p_n \binom{n}{k}$.

Distribution	$\sum_{n=k}^{\infty} p_n \binom{n}{k}$
Poisson	$\dfrac{\theta^k}{(1-e^{-\theta})k!}$
Logarithmic	$-\dfrac{1}{\log(1-\theta)^k}\left(\dfrac{\theta}{1-\theta}\right)^k$
Geometric	$\dfrac{1}{1-\theta}\left(\dfrac{\theta}{1-\theta}\right)^{k-1}$
Binomial	$\dfrac{\theta^k(1+\theta)^{m-k}}{(1+\theta)^m-1}\binom{m}{k}$

symbolic computations such as MAPLE, MATLAB or MATHEMATICA. Hence, several properties of the GPS models can obtained from the exponentiated baseline model like the ordinary and incomplete moments and mgf.

For the sake of illustration, we now provide expansions for the density functions of two well-known GPS distributions:

- An expansion for the WP density function (Lu and Shi, 2012 [214]) is

$$f(x;\theta,\boldsymbol{\tau}) = \sum_{n=1}^{\infty} \frac{\theta^n}{n!(e^\theta-1)} \underbrace{n\,\alpha\,\gamma\,x^{\gamma-1}\exp(-n\,\alpha\,x^\gamma)}_{=\,g_{T_{(1)}}(x;\boldsymbol{\tau})},$$

where $g_{T_{(1)}}(x,\alpha,\beta,n)$ is the density function of $T_{(1)} = \min\{T_i\}_{i=1}^{n}$, for fixed n, with T_i having a *Weibull* distribution. On the other hand, the WP density function can also be expressed as

$$f(x;\theta,\boldsymbol{\tau}) = \sum_{k=1}^{\infty} \nu_k \underbrace{\gamma k\,\alpha^\gamma x^{\gamma-1}\exp\left\{-(\alpha x)^\gamma\right\}\{1-\exp[-(\alpha x)^\gamma]\}^{k-1}}_{=\,g_k(x;\boldsymbol{\tau})}, \quad x>0,$$

where $g_k(x;\boldsymbol{\tau})$ is the EW density function given by Mudholkar and Srivastava (1993) [7].

- An expansion of the BSPS pdf (Bourguignon *et al.*, 2013 [221]) is

$$f(x;\theta,\boldsymbol{\tau}) = \sum_{n=1}^{\infty} p_n \underbrace{n\,\kappa(\alpha,\beta)\,t^{-3/2}\,(t+\beta)\exp\left[-\frac{\zeta(t/\beta)}{2\alpha^2}\right][1-\Phi(v)]^{n-1}}_{=\,g_{T_{(1)}}(x;\boldsymbol{\tau})},$$

for $x > 0$, where $\kappa(\alpha, \beta) = \exp(\alpha^{-2})/(2\alpha\sqrt{2\pi\beta})$ and $\zeta(z) = z + z^{-1}$. Further, $\upsilon - \alpha^{-1}\rho(t/\beta)$, $\rho(z) - z^{1/2} - z^{-1/2}$ and $\Phi(\cdot)$ is the standard normal cdf. Here, $g_{T_{(1)}}(x; \alpha, \beta, n)$ is the density function of $T_{(1)} = \min\{T_i\}_{i=1}^{n}$, for fixed n, where T_i has a *Birnbaum-Saunders* distribution.

On the other hand, the density function of Y defined by (8.4) can be expressed as a mixture of densities of maximum order statistics of the baseline random variable T as follows

$$f(y; \theta, \boldsymbol{\tau}) = \sum_{n=1}^{\infty} p_n \, g_{T_{(n)}}(y; \boldsymbol{\tau}), \quad y > 0, \tag{8.12}$$

where $g_{T_{(n)}}(\cdot)$ is the density function of $T_{(n)} = \max\{T_i\}_{i=1}^{n}$, for fixed n, say

$$g_{T_{(n)}}(t; \boldsymbol{\tau}) = n \, g(t; \boldsymbol{\tau}) G(t; \boldsymbol{\tau})^{n-1}, \quad t > 0.$$

For example, the CEWPS distributions (Cordeiro and Silva, 2015 [222]) can be expressed as

$$f(y; \theta, \alpha, \boldsymbol{\tau}) = \sum_{n=1}^{\infty} p_n \underbrace{n \, \alpha \, h(y; \boldsymbol{\tau}) \mathrm{e}^{-\alpha H(y;\boldsymbol{\tau})} [1 - \mathrm{e}^{-\alpha H(y;\boldsymbol{\tau})}]^{n-1}}_{=g_{T_{(n)}}(y; \alpha, \boldsymbol{\tau})}, \quad x > 0.$$

Since $|\mathrm{e}^{-\alpha H(y;\boldsymbol{\tau})}| < 1$, the quantity $J = [1 - \mathrm{e}^{-\alpha H(y;\boldsymbol{\tau})}]^{n-1}$ can be expanded by the binomial theorem as

$$J = \sum_{j=0}^{n-1}(-1)^j \binom{n-1}{j} \mathrm{e}^{-j\alpha H(y;\boldsymbol{\tau})},$$

which gives

$$f(y; \theta, \alpha, \boldsymbol{\tau}) = \sum_{n=1}^{\infty} \sum_{j=0}^{n-1} \omega_{n,j} \, g(y; (j+1)\alpha, \boldsymbol{\tau}), \quad x > 0,$$

where $\omega_{n,j} = \frac{(-1)^j}{(j+1)} \binom{n-1}{j} p_n$ and $g(y; (j+1)\alpha, \boldsymbol{\tau})$ is given by (8.6).

8.4. OTHER QUANTITIES

Let $X \sim \mathrm{GPS}(\theta, \boldsymbol{\tau})$ and $\mu_s' = \mathrm{E}(X^s) = \int_0^{\infty} x^s f(x; \theta, \boldsymbol{\tau}) \mathrm{d}x$ be its sth moment. The moments of GPS models can be determined using numerical integration from their own density functions. Another way to obtain μ_s' follows from equation (8.9) and the monotone convergence theorem as

$$\mu_s' = \sum_{n=1}^{\infty} p_n \, \mathrm{E}[T_{(1)}^s], \tag{8.13}$$

where $T_{(1)}$ has the distribution of the minimum order statistic of $T_1, \ldots, T_n$, for every n. An alternative expression based on equation (8.11) is

$$\mu'_s = \sum_{k=1}^{\infty} \nu_k \, E(Z_k^s),$$

where Z_k is the exponentiated baseline distribution with cdf $G_k(\cdot)$. The central moments (μ_r), cumulants (κ_r) and skewness and kurtosis of X can be determined as discussed in Section 1.5.

The sth incomplete moment and generating function of X follow from (8.9) or (8.11) using the monotone convergence theorem

$$m_{s:X}(\omega) = \sum_{n=1}^{\infty} p_n \, m_{s:T_{(1)}}(\omega) \qquad \text{or} \qquad m_{s:X}(\omega) = \sum_{k=1}^{\infty} \nu_k \, m_{s:Z}(\omega)$$

and

$$M_X(t) = \sum_{n=1}^{\infty} p_n \, E[e^{t\,T_{(1)}}] \qquad \text{or} \qquad M_X(t) = \sum_{k=1}^{\infty} \nu_k \, E(e^{tZ}),$$

respectively, where Z has the exponentiated baseline distribution and $m_{s:W}(\delta) = \int_0^{\delta} w^s \, f_W(w) \mathrm{d}w$ is the incomplete moment function.

In an analogous way, let $Y \sim \mathrm{CGPS}(\theta, \boldsymbol{\tau})$ and $\mu'_s = \mathrm{E}(Y^s) = \int_0^{\infty} y^s f(y; \theta, \boldsymbol{\tau}) \mathrm{d}y$ be its sth ordinary moment. The ordinary moments of the CGPS models can also be determined using numerical integration from their own density functions. However, from equation (8.12), μ'_s can be expressed as

$$\mu'_s = \sum_{n=1}^{\infty} p_n \, \mathrm{E}[T_{(n)}^s],$$

where $T_{(n)}$ has the distribution of the maximum order statistic of $T_1, \ldots, T_n$, for every n. Hence, the incomplete moments and generating function of Y are

$$m_{s:Y}(\omega) = \sum_{n=1}^{\infty} p_n \, m_{s:T(n)}(\omega) \qquad \text{and} \qquad M_Y(t) = \sum_{n=1}^{\infty} p_n \, E[e^{t\,T_{(n)}}].$$

For the sake of illustration, given N, consider a sequence $T_1, \ldots, T_N$ of iid random variables having the *Weibull* distribution with parameters $\alpha, \beta > 0$. Equation $\mathrm{E}(T^s) = \beta^{-s} \Gamma(s/\alpha + 1)$ implies $\mathrm{E}[T_{(1)}^s] = (N^{1/\alpha}\beta)^{-s}\Gamma(s/\alpha + 1)$ for fixed N. If N is a discrete random variable as anyone in Table 8.1, then $X = \min\{T_i\}_{i=1}^{N}$ has a $\mathrm{WPS}(\theta, \alpha, \beta)$ distribution (Morais and Barreto-Souza, 2011 [215]) with density function

$$f(x; \theta, \alpha, \beta) = \frac{\alpha\,\theta\,\beta^{\alpha}}{C(\theta)} x^{\alpha-1} \mathrm{e}^{-(\beta x)^{\alpha}} C'[\theta \mathrm{e}^{-(\beta x)^{\alpha}}], \quad x > 0.$$

Using $E[T_{(1)}^s]$ in equation (8.13), we obtain

$$E(X^s) = \sum_{n=1}^{\infty} p_n \underbrace{\frac{\Gamma(s/\alpha + 1)}{(n^{1/\alpha}\beta)^s}}_{E[T_{(1)}^s]} = \frac{\Gamma(s/\alpha + 1)}{\beta^s C(\theta)} \sum_{n=1}^{\infty} \frac{a_n \theta^n}{n^{s/\alpha}},$$

where $T_{(1)} \sim W(\alpha, n^{1/\alpha}\beta)$.

For the complementary compounding distributions, consider $Y \sim \text{CEPS}(\theta, \lambda)$ by Flores *et al.* (2013) [232]. In this case, T follows an exponential distribution with parameter λ and

$$g_{T_{(n)}}(t; \lambda) = \sum_{j=1}^{n} (-1)^{j-1} \binom{n}{j} g_{T_j^*}(t; j\lambda), \quad t > 0,$$

where $T_{(n)}$ has the distribution of the maximum order statistics of $T_1, \ldots, T_n$ and T_j^* follows an exponential distribution with parameter $j\lambda$. Hence, we have

$$E(Y^s) = \sum_{n=1}^{\infty} p_n \underbrace{\frac{\Gamma(s+1)}{\lambda^s} \sum_{j=1}^{n} (-1)^{j-1} \binom{n}{j} j^{-s}}_{E[T_{(n)}^s]}.$$

8.5. ORDER STATISTICS

The density of the ith order statistic, say $f_{i:n}(x)$, for a random sample $X_1, \ldots, X_n$ from the GPS distribution, is

$$f_{i:n}(x; \theta, \boldsymbol{\tau}) = \frac{n!}{(i-1)!(n-i)!} f(x; \theta, \boldsymbol{\tau}) F(x; \theta, \boldsymbol{\tau})^{i-1} [1 - F(x; \theta, \boldsymbol{\tau})]^{n-i}$$

and then

$$f_{i:n}(x; \theta, \boldsymbol{\tau}) = \frac{n! \, f(x; \theta, \boldsymbol{\tau})}{(i-1)!(n-i)!} \left\{ 1 - \frac{C\{\theta[1 - G(x; \boldsymbol{\tau})]\}}{C(\theta)} \right\}^{i-1}$$

$$\times \left\{ \frac{C\{\theta[1 - G(x; \boldsymbol{\tau})]\}}{C(\theta)} \right\}^{n-i}, \quad x > 0, \tag{8.14}$$

where $f(\cdot)$ and $F(\cdot)$ are the density and cumulative functions of X, respectively. By using binomial theorem, we can write (8.14) as (for $x > 0$)

$$
f_{i:n}(x; \theta, \boldsymbol{\tau}) = \frac{n! f(x; \theta, \boldsymbol{\tau})}{(i-1)!(n-i)!} \left\{ \frac{C\left\{\theta[1 - G(x; \boldsymbol{\tau})]\right\}}{C(\theta)} \right\}^{n-i}
$$
$$
\times \sum_{j=0}^{i-1} (-1)^j \binom{i-1}{j} \left\{ \frac{C\left\{\theta[1 - G(x; \boldsymbol{\tau})]\right\}}{C(\theta)} \right\}^{j}
$$
$$
= \frac{n! f(x; \theta, \boldsymbol{\tau})}{(i-1)!(n-i)!} \sum_{j=0}^{i-1} (-1)^j \binom{i-1}{j} \left\{ \frac{C\left\{\theta[1 - G(x; \boldsymbol{\tau})]\right\}}{C(\theta)} \right\}^{n+j-i}
$$
$$
= \frac{n! f(x; \theta, \boldsymbol{\tau})}{(i-1)!(n-i)!} \sum_{j=0}^{i-1} (-1)^j \binom{i-1}{j} S(x; \theta, \boldsymbol{\tau})^{n+j-i},
$$

where $S(x; \theta, \boldsymbol{\tau}) = 1 - F(x; \theta, \boldsymbol{\tau})$ is the survival function of X. The corresponding cumulative function, denoted by $F_{i:n}(x; \theta, \boldsymbol{\tau})$, becomes (for $x > 0$)

$$
F_{i:n}(x; \theta, \boldsymbol{\tau}) = \frac{n!}{(i-1)!(n-i)!} \sum_{j=0}^{n-i} \frac{(-1)^j}{j+i} \binom{n-i}{j} \left\{ 1 - \frac{C\left\{\theta[1 - G(x; \boldsymbol{\tau})]\right\}}{C(\theta)} \right\}^{j+i}
$$
$$
= \frac{n!}{(i-1)!(n-i)!} \sum_{j=0}^{n-i} \frac{(-1)^j}{j+i} \binom{n-i}{j} F(x; \theta, \boldsymbol{\tau})^{j+i}. \tag{8.15}
$$

Using a result by Barakat and Abdelkader (2004) [235], the rth moment of the order statistics $X_{1:n}, \ldots, X_{n:n}$ with cdf (8.15) can be provided:

$$
\mathrm{E}(X_{i:n}^r) = r \sum_{j=n+1-i}^{n} (-1)^{j-n+i-1} \binom{j-1}{n-i} \binom{n}{j} \int_0^\infty x^{r-1} S(x, \theta, \boldsymbol{\tau})^j \mathrm{d}x
$$
$$
= r \sum_{j=n+1-i}^{n} \frac{(-1)^{j-n+i-1}}{C(\theta)^j} \binom{j-1}{n-i} \binom{n}{j} \int_0^\infty x^{r-1} C\left\{\theta[1 - G(x; \boldsymbol{\tau})]\right\}^j \mathrm{d}x,
$$
$$
\tag{8.16}
$$

for $i = 1, \ldots, n$, where $S(\cdot)$ is the survival function defined before.
In the same way, the pdf of the ith order statistic, say $f_{i:n}(y)$, for a random sample $Y_1, \ldots, Y_n$ from the CGPS distribution, is (for $y > 0$)

$$
f_{i:n}(y; \theta, \boldsymbol{\tau}) = \frac{n! f(y; \theta, \boldsymbol{\tau})}{(i-1)!(n-i)!} \left\{ \frac{C[\theta G(y; \boldsymbol{\tau})]}{C(\theta)} \right\}^{i-1} \left\{ 1 - \frac{C[\theta G(y; \boldsymbol{\tau})]}{C(\theta)} \right\}^{n-i}, \quad y > 0. \tag{8.17}
$$

An alternative expression for (8 17) is given by using the binomial theorem as

$$f_{i:n}(y; \theta, \boldsymbol{\tau})$$

$$= \frac{n! \, f(y; \theta, \boldsymbol{\tau})}{(i-1)!(n-i)!} \left\{ \frac{C[\theta \, G(y; \boldsymbol{\tau})]}{C(\theta)} \right\}^{i-1} \sum_{j=0}^{n-i} (-1)^j \binom{n-i}{j} \left\{ \frac{C[\theta \, G(y; \boldsymbol{\tau})]}{C(\theta)} \right\}^j$$

$$= \frac{n! \, f(y; \theta, \boldsymbol{\tau})}{(i-1)!(n-i)!} \sum_{j=0}^{n-i} (-1)^j \binom{n-i}{j} \left\{ \frac{C[\theta \, G(y; \boldsymbol{\tau})]}{C(\theta)} \right\}^{j+i-1}$$

$$= \frac{n! \, f(y; \theta, \boldsymbol{\tau})}{(i-1)!(n-i)!} \sum_{j=0}^{n-i} (-1)^j \binom{n-i}{j} F(y; \theta, \boldsymbol{\tau})^{j+i-1}, \quad y > 0.$$

The corresponding cdf becomes

$$F_{i:n}(x; \theta, \boldsymbol{\tau}) = \frac{n!}{(i-1)!(n-i)!} \sum_{j=0}^{n-i} \frac{(-1)^j}{j+i} \binom{n-i}{j} F(y; \theta, \boldsymbol{\tau})^{j+i}$$

$$= \frac{n!}{(i-1)!(n-i)!} \sum_{j=0}^{n-i} \frac{(-1)^j}{j+i} \binom{n-i}{j} \left\{ \frac{C[\theta \, G(y; \boldsymbol{\tau})]}{C(\theta)} \right\}^{j+i}, \quad y > 0.$$

In an analogous way of equation (8.16), we have

$$\mathrm{E}(Y_{i:n}^r) = r \sum_{j=n+1-i}^{n} \frac{(-1)^{j-n+i-1}}{C(\theta)^j} \binom{j-1}{n-i} \binom{n}{j} \int_0^\infty x^{r-1} \left\{ 1 - \frac{C[\theta \, G(y; \boldsymbol{\tau})]}{C(\theta)} \right\}^j \mathrm{d}y,$$

for $i = 1, \ldots, n$.

For example, consider the random variable $X \sim \mathrm{BSPS}(\theta, \alpha, \beta)$. The density function of $X_{i:n}$ is (Bourguignon *et al.*, 2013 [221]) (for $x > 0$)

$$f_{i:n}(x; \theta, \alpha, \beta) = \frac{n! f(x; \theta, \alpha, \beta)}{(i-1)!(n-i)!} \sum_{j=0}^{i-1} (-1)^j \binom{i-1}{j} \left\{ \frac{C[\theta(1 - \Phi(v))]}{C(\theta)} \right\}^{n+j-i},$$

where $f(x; \theta, \alpha, \beta)$ is the BSPS density function and $\Phi(\cdot)$ and v are defined before. For another example, consider the random variable $Y \sim \mathrm{CEPS}(\theta, \beta)$. Define the ith order statistic by $Y_{i:n}$. Then, the density function of $Y_{i:n}$ is given by (Flores *et al.*, 2011 [232]) (for $y > 0$)

$$f_{i:n}(y; \theta, \beta) = \frac{n! f(y; \theta, \alpha, \beta)}{(i-1)!(n-i)!} \sum_{j=0}^{i-1} (-1)^j \binom{i-1}{j} \left\{ \frac{C[\theta(1 - e^{-\beta y})]}{C(\theta)} \right\}^{n+j-i},$$

where $f(y; \theta, \alpha, \beta)$ is the CEPS density function.

8.6.   ESTIMATION

We obtain the MLEs of the parameters of the GPS family of distributions only from complete samples. Let $x_1, \ldots, x_n$ be observed values from the GPS family with parameters θ and $\boldsymbol{\tau}$. Let $\boldsymbol{\theta} = (\theta, \boldsymbol{\tau}^{\top})^{\top}$ be the $p \times 1$ parameter vector. The total log-likelihood function for $\boldsymbol{\theta}$, say $\ell = \ell(\boldsymbol{\theta})$ is

$$\ell = n[\log \theta - \log C(\theta)] + \sum_{i=1}^{n} \log g(x_i; \boldsymbol{\tau}) + \sum_{i=1}^{n} \log C'\{\theta[1 - G(x_i; \boldsymbol{\tau})]\}. \qquad (8.18)$$

The components of the score function $U_n(\boldsymbol{\theta}) = (\partial \ell_n / \partial \theta, \partial \ell_n / \partial \boldsymbol{\tau})^{\top}$ are

$$\frac{\partial \ell}{\partial \theta} = \frac{n}{\theta} - n \frac{C'(\theta)}{C(\theta)} + \sum_{i=1}^{n}[1 - G(x_i; \boldsymbol{\tau})] \frac{C''\{\theta[1 - G(x_i; \boldsymbol{\tau})]\}}{C'\{\theta[1 - G(x_i; \boldsymbol{\tau})]\}}$$

and

$$\frac{\partial \ell}{\partial \boldsymbol{\tau}} = \sum_{i=1}^{n} \frac{1}{g(x_i; \boldsymbol{\tau})} \frac{\mathrm{d}g(x_i; \boldsymbol{\tau})}{\mathrm{d}\boldsymbol{\tau}} - \theta \sum_{i=1}^{n} \frac{C''\{\theta[1 - G(x_i; \boldsymbol{\tau})]\}}{C'\{\theta[1 - G(x_i; \boldsymbol{\tau})]\}} \frac{\mathrm{d}G(x_i; \boldsymbol{\tau})}{\mathrm{d}\boldsymbol{\tau}}.$$

For interval estimation on the model parameters, we require the observed information matrix

$$J(\boldsymbol{\theta}) = - \begin{pmatrix} U_{\theta\theta} & | & U_{\theta\boldsymbol{\tau}}^{\top} \\ -- & -- & -- \\ U_{\theta\boldsymbol{\tau}} & | & U_{\boldsymbol{\tau}\boldsymbol{\tau}}^{\top} \end{pmatrix},$$

whose elements are given by

$$\frac{\partial^2 \ell}{\partial \theta^2} = -\frac{n}{\theta^2} - n \left\{ \frac{C''(\theta)}{C(\theta)} - \left[\frac{C'(\theta)}{C(\theta)} \right]^2 \right\} + \sum_{i=1}^{n}[1 - G(x_i; \boldsymbol{\tau})]^2 \frac{C'''\{\theta[1 - G(x_i; \boldsymbol{\tau})]\}}{C'\{\theta[1 - G(x_i; \boldsymbol{\tau})]\}}$$

$$- \sum_{i=1}^{n}[1 - G(x_i; \boldsymbol{\tau})]^2 \left[\frac{C''\{\theta[1 - G(x_i; \boldsymbol{\tau})]\}}{C'\{\theta[1 - G(x_i; \boldsymbol{\tau})]\}} \right]^2,$$

$$\frac{\partial^2 \ell}{\partial \theta \partial \boldsymbol{\tau}} = -\sum_{i=1}^{n} \frac{C''\{\theta[1 - G(x_i; \boldsymbol{\tau})]\}}{C'\{\theta[1 - G(x_i; \boldsymbol{\tau})]\}} \frac{\mathrm{d}G(x_i; \boldsymbol{\tau})}{\mathrm{d}\boldsymbol{\tau}}$$

$$- \theta \sum_{i=1}^{n}[1 - G(x_i; \boldsymbol{\tau})] \frac{C'''\{\theta[1 - G(x_i; \boldsymbol{\tau})]\}}{C'\{\theta[1 - G(x_i; \boldsymbol{\tau})]\}} \frac{\mathrm{d}G(x_i; \boldsymbol{\tau})}{\mathrm{d}\boldsymbol{\tau}}$$

$$+ \theta \sum_{i=1}^{n}[1 - G(x_i; \boldsymbol{\tau})] \left[\frac{C''\{\theta[1 - G(x_i; \boldsymbol{\tau})]\}}{C'\{\theta[1 - G(x_i; \boldsymbol{\tau})]\}} \right]^2 \frac{\mathrm{d}G(x_i; \boldsymbol{\tau})}{\mathrm{d}\boldsymbol{\tau}}$$

and

$$
\begin{aligned}
\frac{\partial^2 \ell}{\partial \boldsymbol{\tau}^2} =\ & \sum_{i=1}^{n} \frac{1}{g(x_i; \boldsymbol{\tau})} \frac{\mathrm{d}^2 g(x_i; \boldsymbol{\tau})}{\mathrm{d}\boldsymbol{\tau}^2} - \sum_{i=1}^{n} \frac{1}{g(x_i; \boldsymbol{\tau})} \left[\frac{\mathrm{d}g(x_i; \boldsymbol{\tau})}{\mathrm{d}\boldsymbol{\tau}} \right]^2 \\
& + \theta^2 \sum_{i=1}^{n} \frac{C''' \{\theta[1 - G(x_i; \boldsymbol{\tau})]\}}{C' \{\theta[1 - G(x_i; \boldsymbol{\tau})]\}} \left[\frac{\mathrm{d}G(x_i; \boldsymbol{\tau})}{\mathrm{d}\boldsymbol{\tau}} \right]^2 \\
& - \theta^2 \sum_{i=1}^{n} \left[\frac{C'' \{\theta[1 - G(x_i; \boldsymbol{\tau})]\}}{C' \{\theta[1 - G(x_i; \boldsymbol{\tau})]\}} \right]^2 \left[\frac{\mathrm{d}G(x_i; \boldsymbol{\tau})}{\mathrm{d}\boldsymbol{\tau}} \right]^2 \\
& - \theta \sum_{i=1}^{n} \frac{C'' \{\theta[1 - G(x_i; \boldsymbol{\tau})]\}}{C' \{\theta[1 - G(x_i; \boldsymbol{\tau})]\}} \frac{\mathrm{d}^2 G(x_i; \boldsymbol{\tau})}{\mathrm{d}\boldsymbol{\tau}^2}.
\end{aligned}
$$

Let $\widehat{\boldsymbol{\theta}}$ be the MLE of $\boldsymbol{\theta}$. Under standard regular conditions (Cox and Hinkley, 1974 [236]), we can approximate the distribution of $\sqrt{n}(\widehat{\boldsymbol{\theta}} - \boldsymbol{\theta})$ by the multivariate normal $N_p(0, K(\boldsymbol{\theta})^{-1})$ distribution, where $K(\boldsymbol{\theta}) = \lim_{n\to\infty} n^{-1} J(\boldsymbol{\theta})$ is the unit information matrix and p denotes the number of parameters of the compounded distribution.

8.7. APPLICATIONS

As first empirical example, a real data set is used in order to compare the adjustment of the BXIIPS distribution and those of its sub-models, *i.e.*, the BXIIG, BXIIP, WP and BXII distributions. For this first application, the estimation method is approached by maximum likelihood using the `NLMixed` subroutine in `SAS`.

In the first data set, we have 1,519 observations of budget share for fuel expenditure of British households. They were drawn from the 1980-1982 British Family Expenditure Surveys (FES). In Table 8.5, we tabulate the MLEs of the parameters, $-2\ell(\widehat{\boldsymbol{\theta}})$, the K-S, AIC and BIC statistics for the fitted BXIIP, BXIIG, BXII and WP models to the data. Note that the corresponding MLE's standard errors are showed in parentheses.

Roughly, we can say that the BXIIP, BXIIG, BXII and WP models can be used for modelling these data. However, the statistics listed in Table 8.5 show that the BXIIPS distributions outperform the BXII and WP distributions. Plots of the estimated pdfs and cdfs of the fitted competing models to the first data set are displayed in Figure 8.1 endorsing the superiority of the BXIIG model.

As a second example, we compare the fits of the BSPS distributions and those of the main sub-models, *i.e.*, the BSG, BSP and BS distributions. As in the previous case, the parameters are estimated by maximum likelihood using the `NLMixed` subroutine in `SAS`. We shall analyze the fatigue lives (in hours) of 10 bearings of a certain kind. The data are: 152.7, 172.0, 172.5, 173.3, 193.0, 204.7, 216.5, 234.9, 262.6, 422.6. Table 8.6 gives the MLEs (standard errors between parentheses) of the parameters, the values of $-2\ell(\widehat{\boldsymbol{\theta}})$ and of the K-S, AIC and BIC statistics for the fitted BSG,

Table 8.5: Parameter estimates, K-S, AIC and BIC statistics for expenditure data

Distribution	$\hat{\theta}$	$\hat{c}$	$\hat{k}$	$\hat{\alpha}$	$\hat{\beta}$	K–S	$-2\ell(\hat{\Theta})$	AIC	BIC
BXIIP	4.8519	2.3229	49.4268	-	-	0.2216	-5219	-5213	-5197
	(0.4989)	(0.0442)	(7.8702)						
BXIIG	0.9997	3.2964	1.3124	-	-	0.2161	-5276	-5270	-5254
	(0.0001)	(0.0697)	(1.0877)						
BXII	-	1.9063	76.8212	-	-	0.2617	-5082	-5078	-5067
	-	(0.0336)	(5.4854)						
WP	4.9389	-	-	2.4126	47.1347	0.2278	-5215	-5209	-5193
	(0.4968)	-	-	(0.0446)	(7.4834)				

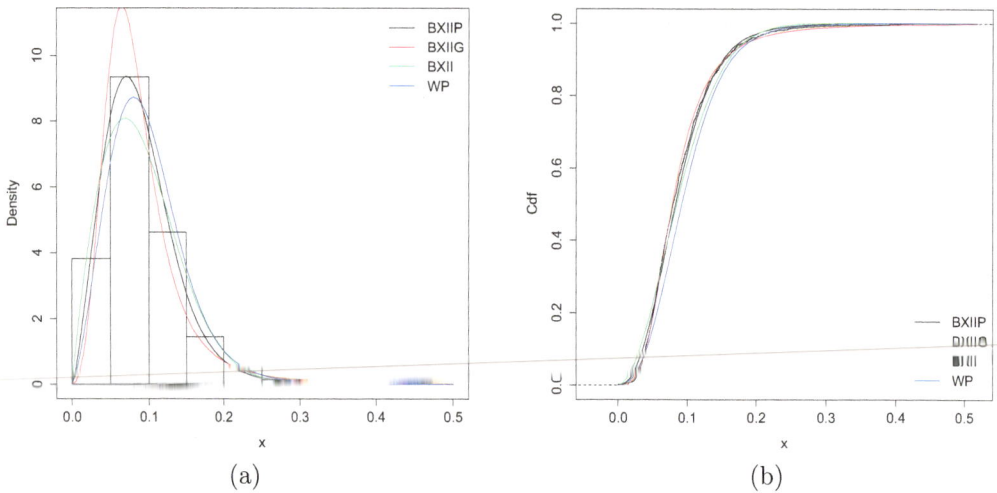

(a)   (b)

Figure 8.1: Estimated (a) pdf and (b) cdf for the BXIIP, BXIIG, BXII and WP models for expenditure data.

BSP and BS models to the second data set. The figures in Table 8.6 indicate that the BSG model yields a better fit to these data than the other models. Plots of the fitted BSG, BSP and BS densities are displayed in Figure 8.2.

Table 8.6: Parameter estimates, K-S statistics, AIC and BIC for fatigue life data

Distribution	$\widehat{\alpha}$	$\widehat{\beta}$	$\widehat{\theta}$	K–S	$-2\ell(\widehat{\Theta})$	AIC	BIC
BSG	0.3087	350.98	0.9672	0.1681	106.9	112.9	113.8
	$(0.1285)^a$	(182.50)	(0.0861)				
BSP	0.2917	259.20	3.1140	0.1633	108.3	114.3	115.2
	$(0.0772)^a$	(44.4148)	(2.4589)				
BS	0.2825	212.05		0.1707	109.9	113.9	114.5
	$(0.0632)^a$	(18.7530)					

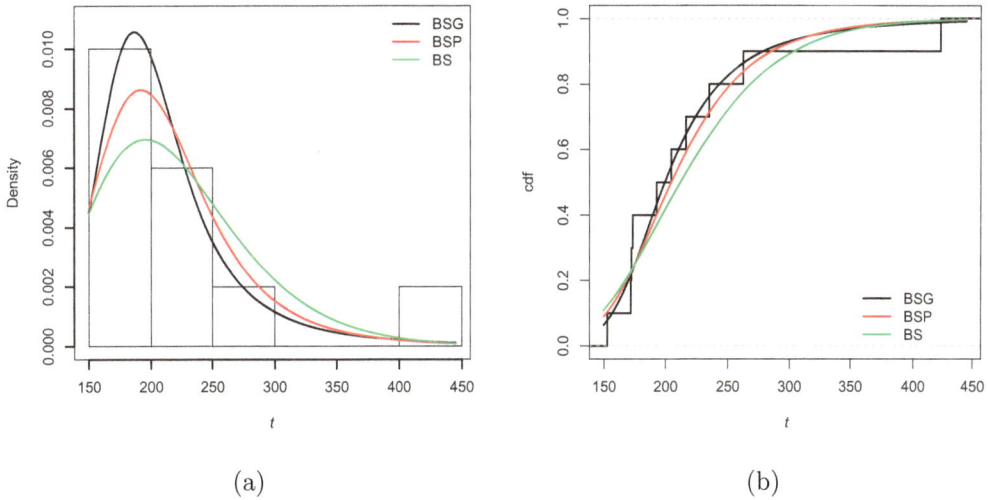

(a) (b)

Figure 8.2: Estimated (a) pdf and (b) cdf for the BSG, BSP and BS models for fatigue life data.

8.8. CONCLUSIONS

In this chapter, we study a wider family of compound models, which includes as special cases the baseline distributions. For any parent continuous distribution G, we can define the corresponding G-power series (GPS) distribution with an extra parameter θ. So, the new family extends several common lifetime distributions such as the exponential, Weibull, modified Weibull, Birnbaum-Saunders and Burr XII distributions. The mathematical properties of the wider family such as the ordinary and incomplete moments, generating and quantile functions and order statistics are determinated for any GPS distribution. The model parameters are estimated by the maximum likelihood method. Two applications to real data sets are provided to prove empirically the flexibility of the compounding models.

Chapter 9

Conclusions and Recent Advances

In recent years, the proposal of new distributions by addying one or more shape parameter(s) to baseline models has received considerable attention. These extended models have been defined as special distribution generators. This book provides the main current generators. According to Tahir and Nadarajah (2015) [160], these generators have been sought to describe applications in Engineering, Economics and Demographic by three reasons, among others:

- The tail properties of new distributions can easily be explored by structure of the new generators;

- The adding parameters inducted from the baseline G models have presented evidence to improve the fit quality;

- The refinement of symbolic and numerical computation softwares has allowed to derive some important mathematical and statistical properties.

Besides the five generators discussed in previous chapters, several other have been proposed in recent years. In what follows, we briefly present other generators as well as a recent generator family proposed by Alzaatreh *et al.* (2013) [237], called the *T-X family*. In all cases, $G(x)$ is the baseline cdf and $g(x) = \mathrm{d}G(x)/\mathrm{d}x$ is its associated pdf. Unlike other chapters, this presentation does not have the aim of discussing mathematical properties; on the other hand, its aim is twofold: (i) to compact the presented generators by means of the T-X family as well as (ii) to indicate other new generators following the proposal of Alzaatreh *et al.* (2013) [237]. Table 9.1 provides with the main current generators.

Marshall and Olkin (1997) [4] derived an important method of including an extra parameter to a given baseline model. Besides to provide an extended distribution, the Marshall and Olkin ("MO" for short) transformation furnishes a wide range of behaviours with respect to the baseline distribution. Santos-Neto *et al.* (2014) [238]

introduced the Marshall-Olkin extended Weibull family, which includes as special cases several models studied in the literature such as the Marshall-Olkin Weibull, Marshall-Olkin Lomax, Marshal-Olkin Fréchet and Marshall-Olkin Burr XII distributions, among others. They defined at least twenty-one special models. Considering the proportion odds model, Sankaran and Jayakumar (2006) [239] presented a detailed discussion about the physical interpretation of the MO family. This family has a relationship with the odds ratio associated with the baseline distribution. Let X be a distributed MO random variable, which describes the lifetime relative to each individual in the population with a vector of p-covariates $\boldsymbol{z} = (z_1, \ldots, z_p)^{\top}$. Then, the cdf and sf of X (for $x \in \mathcal{X} \subseteq \mathbb{R}$) are, respectively, given by

$$F(x\,;\boldsymbol{z}) = \frac{\overline{G}(x)}{1 - [1 - k(\boldsymbol{z})]\,\overline{G}(x)} \quad \text{and} \quad \overline{F}(x\,;\boldsymbol{z}) = \frac{k(\boldsymbol{z})\,\overline{G}(x)}{1 - [1 - k(\boldsymbol{z})]\,\overline{G}(x)}, \qquad (9.1)$$

where $k(\boldsymbol{z}) = \lambda_G(x)/\lambda_F(x\,;\boldsymbol{z})$ is a non-negative function such that $\boldsymbol{z}$ is independent of the time x, $\lambda_F(x\,;\boldsymbol{z})$ is the proportional odds model [for a discussion about such modeling, see Sankaran and Jayakumar (2006)] and $\lambda_G(x) = G(x)/\overline{G}(x)$ represents an arbitrary odds from the baseline distribution. In general, it is often taken $k(\boldsymbol{z}) = \alpha$. Thus, the generator (9.1) has pdf

$$f(x;\alpha) = \frac{\alpha g(x)}{[1 - \overline{\alpha}\,\overline{G}(x)]^2}, \quad x \in \mathcal{X} \subseteq \mathbb{R},\ \alpha > 0.$$

The MO hrf is

$$h(x;\alpha) = \frac{g(x)}{\overline{G}(x)[1 - \overline{\alpha}\,\overline{G}(x)]}, \quad x \in \mathcal{X} \subseteq \mathbb{R},\ \alpha > 0.$$

Recently, Jayakumar and Mathew (2008) [240] have extended (9.1) as (for $x \in \mathcal{X} \subseteq \mathbb{R}$)

$$F(x\,;\alpha,\theta) = 1 - \left[\frac{\alpha \overline{G}(x)}{1 - \overline{\alpha}\,\overline{G}(x)}\right]^{\theta} \quad \text{and} \quad \overline{F}(x;\alpha,\theta) = \left[\frac{\alpha \overline{G}(x)}{1 - \overline{\alpha}\,\overline{G}(x)}\right]^{\theta}.$$

This case is known as the *generalized Marshall-Olkin* generator.

Following the same idea of Eugene *et al.* [100], Alexander *et al.* [86] introduced the generalized beta-generated (GBG) distribution, which has as sub-models the classical beta-generated, Kumaraswamy-generated and exponentiated families. In this case, instead of using the original beta model, the last authors based on the generalized beta distribution of the first kind introduced by McDonald [241] with density

$$f(x) = \frac{c}{B\left(\frac{a}{c}, b\right)}\, x^{a/c-1}\,(1 - x^c)^{b-1},$$

where $a > 0$, $b > 0$ and $c > 0$ are shape parameters. Based on an arbitrary baseline continuous distribution function $G(x)$, the cdf $F(x)$ of the GBG family can be expressed as (for $x \in \mathcal{X} \subseteq \mathbb{R}$)

$$F(x) = I_{G(x)^c}\left(\frac{a}{c}, b\right) = \frac{1}{B\left(\frac{a}{c}, b\right)} \int_0^{G(x)^c} \omega^{a/c-1}(1 - \omega)^{b-1}\mathrm{d}\omega. \qquad (9.2)$$

This case is denoted as *Mc-G* generator. The pdf (for $x \in \mathcal{X} \subseteq \mathbb{R}$) corresponding to (9.2) is

$$f(x; a, b, c) = \frac{c}{B(a, b)} \, g(x) \, G(x)^{ac-1} \left[1 - G(x)^c\right]^{b-1},$$

where $a, b, c > 0$. The Mc hrf is

$$h(x) = \frac{c \, g(x) \, G(x)^{ac-1} \left[1 - G(x)^c\right]^{b-1}}{B_{G(x)^c}(a, b)}.$$

Table 9.1: Summary of main generators of distributions

Short name	Generators by authors
MO	Marshal-Olkin family by Marshall and Olkin (1997) [4]
beta-G	beta-G by Eugene *et al.* (2002) [100] and Jones (2004) [101]
Kw-G	Kumaraswamy-G (Kw-G) by Cordeiro and de Castro (2011) [87]
Mc-G	McDonald-G (Mc-G) by Alexander *et al.* (2012) [86]
gamma-X	gamma-X by Alzaatreh *et al.* (2013a) [242]
gamma-G (type 1)	gamma-G (type 1) by Zografos and Balakrishanan (2009) [201]
gamma-G (type 2)	gamma-G (type 2) by Ristić and Balakrishanan (2012) [205]
gamma-G (type 3)	gamma-G (type 3) by Torabi and Montazari (2012) [243]
log-gamma-G	log-gamma-G by Amini *et al.* (2012) [244]
logistic-G	logistic-G by Torabi and Montazari (2013) [245]
exponentiated generalized-G	exponentiated generalized-G by Cordeiro *et al.* (2013) [246]
T-X	transformed-transformer (T-X) by Alzaatreh *et al.* (2013a) [242]
exponentiated T-X	exponentiated T-X by Alzaghal *et al.* (2013) [247]
Weibull-G	Weibull-G by Bourguignon *et al.* (2014) [248]
exponentiated half-logistic-G	exponentiated half-logistic-G by Cordeiro *et al.* (2014) [249]

Next, we consider the definitions of the cdf and pdf of the T-X family. Let $r(t)$ be the pdf of a random variable $T \in [a, b]$ for $-\infty < a < b < \infty$ and let $W[G(x)]$ be a cdf of a random variable X such that $W[G(x)]$ satisfies the following conditions:

$$\begin{cases} (i) \ W[G(x)] \in [a, b], \\ (ii) \ W[G(x)] \text{ is differentiable and monotonically non-decreasing, and} \\ (iii) \ W[G(x)] \to a \ \text{as} \ x \to -\infty \ \text{and} \ W[G(x)] \to b \ \text{as} \ x \to \infty. \end{cases} \quad (9.3)$$

Recently, Alzaatreh *et al.* (2013) [237] defined the *T-X family* cdf as

$$F(x) = \int_a^{W[G(x)]} r(t) \, dt, \quad (9.4)$$

where $W[G(x)]$ satisfies the conditions (9.3). The pdf corresponding to (9.4) has the form

$$f(x) = \left\{ \frac{d}{dx} W[G(x)] \right\} r\{ W[G(x)] \}. \quad (9.5)$$

New families can be defined from (9.5) by giving the function $W(.)$ such as $W(x) = -\log(1-x), x/(1-x), \log(x/1-x), \log[-\log(x)]$ for $x \in (0,1)$. In Table 9.2, we provide cases of the function $W[G(x)]$ for special models of the T-X family.

Table 9.2: $W[G(x)]$ functions for special T-X models

S.No.	W[G(x)]	Range of T	T-X model
1	$G(x)$	$[0,1]$	Beta-G [100]
			Kw-G type 1 [87]
			Mc-G [86]
			Exp-G (Kw-G type 2) [246]
2	$-\log[G(x)]$	$(0,\infty)$	Gamma-G Type-2 [205]
			Log-Gamma-G Type-2 [244]
			New Weibull-G [250]
3	$-\log[1-G(x)]$	$(0,\infty)$	Gamma-G Type-1 [201]
			Log-Gamma-G Type-1 [244]
			Weibull-X [242, 237]
			Gamma-X [251, 242, 252]
			Exponentiated Half-Logistic-G [249]
4	$-\log[1-G^\alpha(x)]$	$(0,\infty)$	Exponentiated T-X [247]
			Exponentiated Weibull-X [247]
			Exponentiated Gamma-X [247]
5	$G(x)/[1-G(x)]$	$(0,\infty)$	Gamma-G Type-3 [243]
			Weibull-G [248]
6	$\log\{G(x)/[1-G(x)]\}$	$(-\infty,\infty)$	Logistic-G [245]
7	$\log\{-\log[1-G(x)]\}$	$(-\infty,\infty)$	Logistic-X [253]

Finally, we present a family based on the Lomax distribution. The Lomax-G (Cordeiro *et al.*, 2014 [254]) generator with two extra positive parameters α and β is defined by the cdf and pdf

$$F(x) = \alpha\beta^\alpha \int_0^{-\log[1-G(x)]} \frac{dt}{(\beta+t)^{\alpha+1}} = 1 - \left\{\frac{\beta}{\beta-\log[1-G(x)]}\right\}^\alpha$$

and

$$f(x) = \alpha\beta^\alpha \frac{g(x)}{[1-G(x)]\{\beta-\log[1-G(x)]\}},$$

respectively, where $g(x) = dG(x)/dx$. The Lomax-G generator has the same parameters of the baseline G distribution plus two additional parameters α and β.

Bibliography

[1] C. D. Lai, "Constructions and applications of lifetime distributions," *Appl. Stoch. Model Bus.*, vol. 29, pp. 127–140, 2013.

[2] M. E. Ghitany, F. A. Al-Awadhi, and L. A. Alkhalfan, "Marshall-Olkin extended Lomax distribution and its application," *Commun. Stat.-Theory Methods*, vol. 36, pp. 1855–1866, 2007.

[3] W. Q. Meeker and L. A. Escobar, *Statistical Methods for Reliability Data.* New York: John Wiley & Sons, 1998.

[4] A. W. Marshall and I. Olkin, "A new method for adding a parameter to a family of distributions with application to the exponential and Weibull families," *Biometrika*, vol. 84, pp. 641–652, 1997.

[5] J. M. F. Carrasco, E. M. M. Ortega, and G. M. Cordeiro, "A generalized modified Weibull distribution for lifetime modeling." *Comput. Stat. Data An.*, vol. 53, pp. 450–462, 2008.

[6] C. D. Lai, "Generalized Erlang and mortality levelling off distributions," *Math. Comput. Model.*, vol. 51, pp. 1268–1276, 2010.

[7] G. S. Mudholkar and D. K. Srivastava, "Exponentiated Weibull family for analyzing bathtub failure-rate data," *IEEE Trans. Rel.*, vol. 42, pp. 299–302, 1993.

[8] J. Lawless, *Statistical Models and Methods for Lifetime Data*, ser. Wiley Series in Probability and Statistics.   Wiley, 2011.

[9] R. B. Silva, W. Barreto-Souza, and G. M. Cordeiro, "A new distribution with decreasing, increasing and upside-down bathtub failure rate," *Comput. Stat. Data An.*, vol. 54, pp. 935–944, 2010.

[10] M. Tableman and J. Kim, *Survival Analysis Using S: Analysis of Time-to-Event Data*, ser. Chapman & Hall/CRC Texts in Stat. Sci.   Taylor & Francis, 2003.

[11] I. S. Gradshteyn and I. M. Ryzhik, *Table of Integrals, Series, and Products.* New York: Academic, 2000.

[12] M. Abramowitz and I. Stegun, *Handbook of Mathematical Functions: With Formulas, Graphs, and Mathematical Tables*, ser. Applied mathematics series. Dover Publications, 1965.

[13] H. Exton, *Handbook of Hypergeometric Integrals: Theory, Applications, Tables, Computer Programs.* New York, Halsted Press, 1978.

[14] A. A. A. Kilbas, H. M. Srivastava, and J. J. Trujillo, *Theory And Applications of Fractional Differential Equations*, ser. North-Holland Mathematics Studies. Elsevier Science & Tech, 2006.

[15] J. A. Greenwood, J. M. Landwehr, N. C. Matalas, and J. R. Wallis, "Probability weighted moments: definition and relation to parameters of several distributions expressable in inverse form." *Water Resour. Res.*, vol. 15, pp. 1049–1064, 1979.

[16] J. Hosking, "L-moments: analysis and estimation of distributions using linear combinations of order statistics," *J. R. Stat. Soc. Series B Stat. Methodol.*, vol. 52, pp. 105–124, 1990.

[17] ——, "Distributions with maximum entropy subject to constraints on their l-moments or expected order statistics," *J. Stat. Plan. Inference*, vol. 137, pp. 2870–2891, 2007.

[18] R. V. L. Hartley, "Transmission of information," *Bell Sys. Tech. J.*, vol. 7, pp. 535–563, 1928.

[19] C. E. Shannon, "A mathematical theory of communication," *Bell Sys. Tech. J.*, vol. 27, pp. 379–423, 1948.

[20] S. Kullback and R. A. Leibler, "On information and sufficiency," *Ann. Math. Stat.*, vol. 22, pp. 79–86, 1951.

[21] S. Nadarajah, G. Cordeiro, and E. Ortega, "General results for the beta-modified weibull distribution," *J. Stat. Comput. Simul.*, vol. 81, 2011.

[22] G. Cordeiro, A. J. Lemonte, and E. Ortega, "The Marshall–Olkin family of distributions: Mathematical properties and new models," *J. Stat. Theory Pract.*, vol. 8, 2014.

[23] G. Chen and N. Balakrishnan, "A general purpose approximate goodness-of-fit test," *J. Qual. Technol.*, vol. 27, pp. 154–161, 1995.

[24] R. Pakyari and N. Balakrishnan, "A general purpose approximate goodness-of-fit test for progressively Type-II censored data," *IEEE Trans. Rel.*, vol. 61, pp. 238–244, March 2012.

[25] S. Nadarajah, "The exponentiated exponential distribution: a survey," *AStA Adv. Stat. Anal.*, vol. 95, pp. 219–251, 2011.

[26] S. Nadarajah, G. M. Cordeiro, and E. M. M. Ortega, "The exponentiated Weibull distribution: a survey," *Stat. Pap.*, vol. 54, pp. 839–877, 2013.

[27] P. G. Sankaran and V. L. Gleeja, "On bivariate reversed hazard rates," *J. Japan Stat. Soc.*, vol. 36, pp. 213–224, 2006.

[28] A. M. Variyath and P. G. Sankaran, "Parametric regression models using reversed hazard rates," *J. Probab. Stat.*, vol. 2014, pp. 1–5, 2014.

[29] D. Kundu and R. D. Gupta, "A class of bivariate models with proportional reversed hazard marginals," *Sankhya B*, vol. 72, pp. 236–253, 2010.

[30] E. J. Veres-Ferrer and J. M. Pavía, "On the relationship between the reversed hazard rate and elasticity," *Stat. Pap.*, vol. 22, pp. 1–10, 2012.

[31] E. L. Lehmann, "The power of rank tests," *Ann. Stat.*, vol. 24, pp. 28–43, 1953.

[32] J. Rodrigues, N. Balakrishnan, G. M. Cordeiro, and M. de Castro, "A unified view on lifetime distributions arising from selection mechanisms," *Comput. Stat. Data An.*, vol. 55, pp. 3311–3319, 2011.

[33] W. Barreto-Souza and F. Cribari-Neto, "A generalization of the exponential-Poisson distribution," *Stat. Probab. Lett.*, vol. 79, pp. 2493–2500, 2009.

[34] W. Barreto-Souza, R. B. Silva, and G. M. Cordeiro, "A new distribution with decreasing, increasing and upside-down bathtub failure rate," *Comput. Stat. Data An.*, vol. 54, pp. 935–944, 2010.

[35] M. Chahkandi and M. Ganjali, "On some lifetime distributions with decreasing failure rate," *Comput. Stat. Data An.*, vol. 53, pp. 4433–4440, 2009.

[36] G. S. Mudholkar, D. K. Srivastava, and M. Freimer, "The exponentiated Weibull family: A reanalysis of the bus-motor-failure data," *Technometrics*, vol. 37, pp. 436–445, 1995.

[37] R. C. Gupta, R. D. Gupta, and P. L. Gupta, "Modeling failure time data by Lehmann alternatives," *Commun. Stat.-Theory Methods*, vol. 27, pp. 887–904, 1998.

[38] R. D. Gupta and D. Kundu, "Exponentiated exponential family: An alternative to gamma and Weibull distributions," *Biom. J.*, vol. 43, pp. 117–130, 2001.

[39] S. Nadarajah and A. K. Gupta, "The exponentiated gamma distribution with application to drought data," *Cal. Stat. Assoc. Bull.*, vol. 59, pp. 29–54, 2007.

[40] G. M. Cordeiro, E. M. M. Ortega, and G. O. Silva, "The exponentiated generalized gamma distribution with application to lifetime data," *J. Stat. Comput. Simul.*, vol. 81, pp. 827–842, 2011.

[41] S. Nadarajah and S. Kotz, "The exponentiated type distributions," *Acta Appl. Math.*, vol. 92, pp. 97–111, 2006.

[42] R. D. Gupta and D. Kundu, "Generalized exponential distributions," *Aust. N. Z. J. Stat.*, vol. 41, pp. 173–188, 1999.

[43] ——, "Generalized exponential distribution: Different method of estimations," *J. Stat. Comput. Simul.*, vol. 69, pp. 315–337, 2001.

[44] ——, "Generalized exponential distribution: Existing results and some recent developments," *J. Stat. Plan. Inference*, vol. 137, pp. 3537–3547, 2007, special Issue: In Celebration of the Centennial of The Birth of Samarendra Nath Roy (1906-1964).

[45] M. Z. Raqab and M. Ahsanullah, "Estimation of the location and scale parameters of generalized exponential distribution based on order statistics," *J. Stat. Comput. Simul.*, vol. 69, pp. 109–124, 2001.

[46] M. Z. Raqab, "Inferences for generalized exponential distribution based on record statistics," *J. Stat. Plan. Inference*, vol. 104, pp. 339–350, 2002.

[47] A. M. Sarhan, "Analysis of incomplete, censored data in competing risks models with generalized exponential distributions," *IEEE Trans. Rel.*, vol. 56, pp. 132–138, 2007.

[48] A. H. Abdel-Hamid and E. K. Al-Hussaini, "Estimation in step-stress accelerated life tests for the exponentiated exponential distribution with type I censoring," *Comput. Stat. Data An.*, vol. 53, pp. 1328–1338, 2009.

[49] M. Aslam, D. Kundu, and M. Ahmad, "Time truncated acceptance sampling plans for generalized exponential distribution," *J. Appl. Stat.*, vol. 37, pp. 555–566, 2010.

[50] A. I. Shawky and R. A. Bakoban, "Exponentiated gamma distribution: Different methods of estimations," *J. Appl. Math.*, vol. 2012, pp. 1–23, 2012.

[51] K. Persson and J. Rydén, "Exponentiated Gumbel distribution for estimation of return levels of significant wave height," *J. Environ. Stat.*, vol. 1, pp. 1–12, 2010.

[52] D. Kundu and R. D. Gupta, "Generalized exponential distribution: Bayesian estimations," *Comput. Stat. Data An.*, vol. 52, pp. 1873–1883, 2008.

[53] D. Kundu, R. D. Gupta, and A. Manglick, "Discriminating between the log-normal and generalized exponential distributions," *J. Stat. Plan. Inference*, vol. 127, pp. 213–227, 2005.

[54] E. M. Hashimoto, E. M. M. Ortega, V. G. Cancho, and G. M. Cordeiro, "The log-exponentiated Weibull regression model for interval-censored data," *Comput. Stat. Data An.*, vol. 54, pp. 1017–1035, 2010.

[55] B. Gompertz, "On the nature of the function expressive of the law of human mortality, and on a new mode of determining the value of life contingencies," *Philos. Trans. R. Soc. London*, vol. 115, pp. 513–583, 1825.

[56] P. F. Verhulst, *Deuxieme memoire sur la loi d'accroissement de la population*. Memoires de l'Academie Royale des Sciences, des Lettres et des Beaux-Arts de Belgique, 1847.

[57] D. N. P. Murthy, M. Xie, and R. Jiang, *Weibull Models*. Wiley Series in Probability and Statistics, 2004.

[58] H. Rinne, *The Weibull Distribution: A Handbook*. CRC Press, 2009.

[59] G. S. Mudholkar and A. D. Hutson, "The exponentiated Weibull family: some properties and a flood data application," *Commun. Stat.-Theory Methods*, vol. 25, pp. 3059–3083, 1996.

[60] M. M. Nassar and F. H. Eissa, "On the exponentiated Weibull distribution," *Commun. Stat.-Theory Methods*, vol. 32, pp. 1317–1336, 2003.

[61] M. Pal, M. M. Ali, and J. Woo, "Exponentiated Weibull distribution," *Statistica*, vol. LXVI, pp. 139–147, 2006.

[62] S. Nadarajah and A. K. Gupta, "On the moments of the exponentiated Weibull distribution," *Commun. Stat.-Theory Methods*, vol. 34, pp. 253–256, 2005.

[63] H. A. Sartawi and M. S. Abu-Salih, "Bayes prediction bounds for the Burr type X model," *Commun. Stat.-Theory Methods*, vol. 20, pp. 2307–2330, 1991.

[64] D. Kundu and R. D. Gupta, "Characterizations of the proportional (reversed) hazard class," *Commun. Stat.-Theory Methods*, vol. 38, pp. 3095–3102, 2004.

[65] D. Kundu and M. Raqab, "Generalized Rayleigh distribution: Different methods of estimations," *Comput. Stat. Data An.*, vol. 49, pp. 187–200, 2005.

[66] I. Malinowska and D. Szynal, "On characterization of certain distributions of kth lower (upper) record values," *Appl. Math. Compt.*, vol. 202, pp. 338–347, 2008.

[67] M. Zhou, D. Yang, Y. Wang, and S. . Nadarajah, "Moments of the scaled Burr type X distribution," *J. Comput. Anal. Appl.*, vol. 10, pp. 523–525, 2008.

[68] S. Nadarajah, "Batub-shaped failure rate functions," *Qual. Quant.*, vol. 43, pp. 855–863, 2009.

[69] S. Nadarajah and T. Pogány, "On the characteristic functions for extreme value distributions," *Extremes*, vol. 16, pp. 27–38, 2013.

[70] G. S. Mudholkar and I. C. Sarkar, "A proportional hazards modeling of multisample reliability data," *Commun. Stat.-Theory Methods*, vol. 28, pp. 2079–2101, 1999.

[71] A. W. Gera, "The modified exponentiated-Weibull distribution for lifetime modeling," in *Proceedings of the Annual Reliability and Maintainability Symposium*, 1997, pp. 149–152.

[72] C. D. Lai, M. Xie, and D. N. P. Murthy, "A modified Weibull distribution," *IEEE Trans. Rel.*, vol. 52, pp. 33–37, 2003.

[73] V. G. Cancho and H. Bolfarine, "Modeling the presence of immunes by using the exponentiated-Weibull model," *J. Appl. Stat.*, vol. 28, pp. 659–671, 2001.

[74] H. Jiang, M. Xie, and L. C. Tang, "On the odd Weibull distribution," *J. Risk Reliab.*, vol. 222, pp. 583–594, 2008.

[75] M. Nikulin and F. Haghighi, "A chi-squared test for the generalized power Weibull family for the head-and-neck cancer censored data," *J. Math. Sci.*, vol. 133, pp. 1333–1341, 2006.

[76] ——, "On the power generalized Weibull family: model for cancer censored data," *Metron*, vol. LXVII, pp. 75–86, 2009.

[77] V. G. Voda, "A modified Weibull hazard rate as generator of a generalized Maxwell distribution," *Math. Rep.*, vol. 11, pp. 171–179, 2009.

[78] A. S. Wahed, T. M. Luong, and J. H. Jeong, "A new generalization of Weibull distribution with application to a breast cancer data set," *Stat. Med.*, vol. 28, pp. 2077–2094, 2009.

[79] G. M. Cordeiro, E. Ortega, and G. Silva, "The beta extended Weibull family," *J. Probab. Stat. Sci.*, vol. 10, pp. 15–40, 2012.

[80] F. Famoye, C. Lee, and O. Olumolade, "The beta Weibull distribution," *J. Stat. Theory Appl.*, vol. 4, pp. 121–136, 2005.

[81] C. Lee, F. Famoye, and O. Olumolade, "Beta Weibull distribution: Some properties and applications to censored data," *J. Mod. Appl. Stat.*, vol. 6, pp. 173–186, 2007.

[82] G. M. Cordeiro and S. Nadarajah, "Closed-form expressions for moments of a class of beta generalized distributions," *Brazilian J. Probab. Stat.*, vol. 25, pp. 14–33, 2011.

[83] G. M. Cordeiro, E. M. M. Ortega, and G. O. Silva, "The exponentiated generalized gamma distribution with application to lifetime data," *J. Stat. Comput. Simul.*, vol. 81, pp. 827–842, 2011.

[84] G. M. Cordeiro, E. M. M. Ortega, and S. Nadarajah, "The Kumaraswamy Weibull distribution with application to failure data," *J. Franklin Inst.*, vol. 347, pp. 1399–1429., 2010.

[85] G. O. Silva, E. M. M. Ortega, and G. M. Cordeiro, "The beta modified Weibull distribution," *Lifetime Data Anal.*, vol. 16, pp. 409–430, 2010.

[86] C. Alexander, G. M. Cordeiro, E. M. M. Ortega, and J. M. Sarabia, "Generalized beta-generated distributions," *Comput. Stat. Data An.*, vol. 56, pp. 1880–1897, 2012.

[87] G. M. Cordeiro and M. de Castro, "A new family of generalized distributions," *J. Stat. Comput. Simul.*, vol. 81, pp. 883–898, 2011.

[88] S. Nadarajah, G. M. Cordeiro, and E. M. M. Ortega, "General results for the Kumaraswamy-G distribution," *J. Stat. Comput. Simul.*, vol. 82, pp. 951–979, 2012.

[89] A. J. Lemonte, W. Barreto-Souza, and G. M. Cordeiro, "The exponentiated Kumaraswamy distribution and its log-transform," *Brazilian J. Probab. Stat.*, vol. 27, pp. 31–53, 2011.

[90] G. M. Cordeiro, A. B. Simas, and B. D. Stosic, "Closed form expressions for moments of the beta Weibull distribution," *An. Acad. Bras. Ciênc.*, vol. 83, pp. 357–373, 2011.

[91] M. A. P. Pascoa, E. M. M. Ortega, G. M. Cordeiro, and P. F. Paranaíba, "The Kumaraswamy-generalized gamma distribution with application in survival analysis," *Stat. Methodol.*, vol. 8, pp. 411–433, 2011.

[92] G. M. Giorgi and S. Nadarajah, "Bonferroni and Gini indices for various parametric families of distributions," *Metron*, vol. 68, pp. 23–46, 2010.

[93] J. M. Sarabia and E. Castillo, "About a class of max-stable families with applications to income distributions," *Metron*, vol. LXIII, pp. 505–527, 2005.

[94] M. Z. Raqab and D. Kundu, "Burr type X distribution: Revisited," *J. Probab. Stat. Sci.*, vol. 4, pp. 179–193, 2006.

[95] M. Z. Raqab, "Order statistics from the Burr type X model," *Comput. Math. Appl.*, vol. 36, pp. 111–120, 1998.

[96] R. U. Khan, Z. Anwar, and H. Athar, "Recurrence relations for single and product moments of dual generalized order statistics from exponentiated Weibull distribution," *Aligarh J. Stat.*, vol. 28, pp. 37–45, 2008.

[97] A. Rényi, "On measures of entropy and information," in *4th Berkeley Symposium on Mathematical Statistics and Probability*, vol. 1, 1961, pp. 547–561.

[98] M. Rao, Y. Chen, B. C. Vemuri, and F. Wang, "Cumulative residual entropy: A new measure of information," *IEEE Trans. Inf. Theory*, vol. 50, pp. 1220–1228, 2004.

[99] C. Cheng, J. Chen, and Z. Li, "A new algorithm for MLE of exponentiated Weibull distribution with censoring data." *Appl. Math.*, vol. 23, pp. 638–647, 2010.

[100] N. Eugene, C. Lee, and F. Famoye, "Beta-normal distribution and its applications," *Commun. Stat.-Theory Methods*, vol. 31, pp. 497–512, 2002.

[101] M. C. Jones, "Families of distributions arising from the distributions of order stat." *TEST*, vol. 13, pp. 1–43, 2004.

[102] A. K. Gupta and S. Nadarajah, "On the moments of the beta normal distribution," *Commun. Stat.-Theory Methods*, vol. 33, pp. 1–13, 2004.

[103] M. Razzaghi, "Beta-normal distribution in dose-response modeling and risk assessment for quantitative responses," *Environ. Ecol. Stat.*, vol. 16, pp. 25–36, 2009.

[104] L. C. Rêgo, R. J. Cintra, and G. M. Cordeiro, "On some properties of the beta normal distribution," *Commun. Stat.-Theory Methods*, vol. 41, pp. 3722–3738, 2012.

[105] R. J. Cintra, L. C. Rêgo, G. M. Cordeiro, and A. D. C. Nascimento, "Beta generalized normal distribution with an application for SAR image processing," *Stat.*, vol. 48, pp. 279–294, 2014.

[106] A. K. Gupta and S. Nadarajah, "Beta Bessel distributions," *Int. J. Math. Math. Sci.*, vol. 2006, pp. 1–14, 2006.

[107] J. Rodríguez-Avi, A. Conde-Sánchez, A. J. Sáez-Castillo, and M. J. Olmo-Jiménez, "Gaussian hypergeometric probability distributions for fitting discrete data," *Commun. Stat.-Theory Methods*, vol. 36, pp. 453–463, 2007.

[108] A. Akinsete, F. Famoye, and C. Lee, "The beta-Pareto distribution," *Stat.*, vol. 42, pp. 547–563, 2008.

[109] T. Kozubowski and S. Nadarajah, "The beta-Laplace distribution," *J. Comput. Anal. Appl.*, vol. 10, pp. 305–318, 2008.

[110] R. R. Pescim, C. G. B. Demétrio, G. M. Cordeiro, E. M. M. Ortega, and M. R. Urbano, "The beta generalized half-normal distribution," *Comput. Stat. Data An.*, vol. 54, pp. 945–957, 2010.

[111] W. Barreto-Souza, G. M. Cordeiro, and A. B. Simas, "Some results for beta Fréchet distribution," *Commun. Stat.- Theory Methods*, vol. 40, pp. 798–811, 2011.

[112] E. Mahmoudi, "The beta generalized Pareto distribution with application to lifetime data," *Math. Comput. Simul.*, vol. 81, pp. 2414–2430, 2011.

[113] S. Nadarajah, G. M. Cordeiro, and E. M. M. Ortega, "General results for the beta-modified Weibull distribution," *J. Stat. Comput. Simul.*, vol. 81, pp. 1211–1232, 2011.

[114] G. M. Cordeiro and A. J. Lemonte, "The beta Laplace distribution," *Stat. Probab. Lett.*, vol. 81, pp. 973–982, 2011.

[115] ——, "The beta-half-Cauchy distribution," *J. Probab. Stat.*, vol. 2011, pp. 1–18, 2011.

[116] F. Castellares, L. C. Montenegro, and G. M. Cordeiro, "The beta log-normal distribution," *J. Stat. Comput. Simul.*, vol. 45, pp. 1–26, 2011.

[117] P. F. Paranaíba, E. M. M. Ortega, G. M. Cordeiro, and R. R. Pescim, "The beta Burr XII distribution with application to lifetime data," *Comput. Stat. Data An.*, vol. 55, pp. 1118–1136, 2011.

[118] G. M. Cordeiro and A. J. Lemonte, "The β-Birnbaum-Saunders distribution: An improved distribution for fatigue life modeling," *Comput. Stat. Data An.*, vol. 55, pp. 1445–1461, 2011.

[119] G. Moutinho Cordeiro and R. dos Santos Brito, "The beta power distribution," *Brazilian J. Probab. Stat.*, vol. 26, pp. 88–112, 2012.

[120] G. M. Cordeiro and A. J. Lemonte, "The McDonald inverted beta distribution," *J. Franklin Inst.*, vol. 349, pp. 1174–1197, 2012.

[121] ——, "The McDonald arcsine distribution: a new model to proportional data," *Stat.*, vol. 48, pp. 182–199, 2014.

[122] G. M. Cordeiro, S. Nadarajah, and E. M. M. Ortega, "General results for the beta Weibull distribution," *J. Stat. Comput. Simul.*, vol. 83, pp. 1082–1114, 2013.

[123] G. Cordeiro, F. Castellares, L. C. Montenegro, and M. de Castro, "The beta generalized gamma distribution," *Stat.*, vol. 47, pp. 888–900, 2013.

[124] G. M. Cordeiro, J. S. Nobre, R. R. Pescim, and E. M. M. Ortega, "The beta Moyal: a useful-skew distribution," *Int. J. Res. Rev. Appl. Sci.*, vol. 10, pp. 171–192, 2012.

[125] A. J. Lemonte, "The beta log-logistic distribution," *Brazilian J. Probab. Stat.*, vol. 28, pp. 313–332, 2014.

[126] N. Singla, K. Jain, and S. Sharma, "The beta generalized Weibull distribution: Properties and applications," *Reliab. Eng. Syst. Safe*, vol. 102, pp. 5–15, 2012.

[127] L. M. Zea, R. Silva, M. Bourguignon, A. Santos, and G. Cordeiro, "The beta exponentiated Pareto distribution with application to bladder cancer susceptibility," *Int. J. Stat. Probab.*, vol. 1, pp. 8–19, 2012.

[128] J. Achcar, E. Coelho-Barros, and G. Cordeiro, "Beta generalized distributions and related exponentiated models: A bayesian approach," *Brazilian J. Probab. Stat.*, vol. 27, pp. 1–19, 2013.

[129] G. Cordeiro, G. Silva, and E. Ortega, "The beta-Weibull geometric distribution," *Stat.*, vol. 47, pp. 817–834, 2013.

[130] G. Cordeiro, C. Cristino, E. M. Hashimoto, and E. Ortega, "The beta generalized rayleigh distribution with applications to lifetime data," *Stat. Pap.*, vol. 54, pp. 133–161, 2013.

[131] G. Cordeiro, A. E. Gomes, C. da Silva, and E. Ortega, "The beta exponentiated Weibull distribution," *J. Stat. Comput. Simul.*, vol. 83, pp. 114–138, 2013.

[132] A. E. Gomes, da Silva C.Q., G. Cordeiro, and E. Ortega, "The beta Burr III model for lifetime data," *Brazilian J. Probab. Stat.*, vol. 27, pp. 502–543, 2013.

[133] G. Cordeiro, G. Silva, R. Pescim, and E. Ortega, "General properties for the beta extended half-normal model," *J. Stat. Comput. Simul.*, vol. 84, pp. 881–901, 2014.

[134] A. J. Lemonte and G. Cordeiro, "An extended Lomax distribution," *Stat.*, vol. 47, pp. 800–816, 2013.

[135] O. Shittu and K. Adepoju, "On the beta-Nakagami distribution," *Progress in Applied Mathematics*, vol. 5, pp. 1–10, 2013.

[136] W. Gilchrist, *Statistical Modelling with Quantile Functions.* Chapman and Hall, New York, 2000.

[137] J. Kenney and E. Keeping, *Mathematics of statistics.* Princeton, NJ., 1962.

[138] R. Gupta and R. Gupta, "Analyzing skewed data by power normal model," *TEST*, vol. 17, pp. 197–210, 2008.

[139] E. K. Al-Hussaini, "Inference based on censored samples from exponentiated populations," *TEST*, vol. 19, pp. 487–513, 2010.

[140] F. R. S. Gusmão, E. M. M. Ortega, and G. M. Cordeiro, "The generalized inverse Weibull distribution," *Stat. Pap.*, vol. 52, pp. 591–619, 2011.

[141] A. J. Lemonte and G. Cordeiro, "The exponentiated generalized inverse gaussian distribution," *Stat. Probab. Lett.*, vol. 81, pp. 506–517, 2011.

[142] E. K. Al-Hussaini and M. Hussain, "Interval estimation based on data from exponentiated Burr XII population," *Am. Open J. Stat.*, vol. 1, pp. 33–45, 2011.

[143] A. J. Lemonte, W. Barreto-Souza, and G. Cordeiro, "The exponentiated Kumaraswamy distribution and its log-transform," *Brazilian J. Probab. Stat.*, vol. 27, pp. 31–53, 2013.

[144] A. J. Lemonte, "A new exponential-type distribution with constant, decreasing, increasing, upside-down bathtub and bathtub-shaped failure rate function," *Comput. Stat. Data An.*, vol. 62, pp. 149–170, 2013.

[145] A. P. Prudnikov, Y. A. Brychkov, and O. I. Marichev, *Integrals and Series*, Gordon and Breach, Eds. Amsterdam: Science Publishers, 1986, vol. 3.

[146] S. Nadarajah, "Explicit expressions for moments of order statistics," *Stat. Probab. Lett.*, vol. 78, pp. 196–205, 2008.

[147] ——, "Explicit expressions for moments of t order statistics," *C. R. Acad. Sci.*, vol. 345, pp. 523–526, 2007.

[148] W. Shaw, "Sampling student's t distribution: use of the inverse cumulative distribution function," *J. Comput. Finance*, vol. 9, pp. 37–73, 2006.

[149] N. Ebrahimi, E. Maasoumi, and E. Soofi, "Ordering univariate distributions by entropy and variance," *J. Econom.*, vol. 90, pp. 317–336, 1999.

[150] J. Doornik, *Ox 5: object-oriented matrix programming language.* Timberlake Consultants, London, 2007.

[151] E. Ortega, G. Cordeiro, and E. M. Hashimoto, "A log-linear regression model for the beta-Weibull distribution," *Commun. Stat.-Simul. Comput.*, vol. 40, pp. 1206–1235, 2011.

[152] E. M. Hashimoto, G. Cordeiro, and E. Ortega, "The new Neyman type a beta Weibull model with long-term survivors," *Comput. Stat.*, vol. 28, pp. 933–954, 2013.

[153] E. Ortega, G. Cordeiro, and M. Kattan, "The negative binomial-beta Weibull regression model to predict the cure of prostate cancer," *J. Appl. Stat.*, vol. 39, pp. 1191–1210, 2012.

[154] E. Ortega, G. Cordeiro, and A. J. Lemonte, "A log-linear regression model for the β-Birnbaum-Saunders distribution with censored data," *Comput. Stat. Data An.*, vol. 56, pp. 698–718, 2012.

[155] E. M. M. Ortega, G. M. Cordeiro, and M. W. Kattan, "The log-beta Weibull regression model with application to predict recurrence of prostate cancer," *Stat. Pap.*, vol. 54, pp. 113–132, 2013.

[156] M. Xie, Y. Tang, and T. N. Goh, "A modified Weibull extension with bathtub failure rate function," *Reliab. Eng. Syst. Safe*, vol. 76, pp. 279–285., 2002.

[157] M. V. Aarset, "How to identify a bathtub hazard rate," *IEEE Trans. Rel.*, vol. 36, pp. 106–108, 1987.

[158] S. Nadarajah and S. Kotz, "The beta Gumbel distribution," *Math. Probl. Eng.*, vol. 2004, pp. 323–332, 2004.

[159] S. Nadarajah and A. K. Gupta, "The beta Fréchet distribution," *Far East J. Theor. Stat.*, vol. 14, pp. 15–24, 2004.

[160] M. Tahir and S. Nadarajah, "Parameter induction in continuous univariate distributions: Well-established G families," *An. Acad. Bras. Ciênc.*, vol. 87, pp. 539–568, 2015.

[161] P. Kumaraswamy, "A generalized probability density function for double-bounded random processes," *J. Hydrol.*, vol. 46, pp. 79–88, 1980.

[162] S. G. Fletcher and K. Ponnambalam, "A new formulation for the stochastic control of systems with bounded state variables: An application to a single reservoir system," *Stoch. Hydrol. Hydraul.*, vol. 10, pp. 167–186, 1996.

[163] A. Seifi, K. Ponnambalam, and J. Vlach, "Maximization of manufacturing yield of systems with arbitrary distributions of component values," *Ann. Oper. Res.*, vol. 99, pp. 373–383, 2000.

[164] M. C. Jones, "Kumaraswamy's distribution: A beta-type distribution with some tractability advantages," *Stat. Methodol.*, vol. 6, pp. 70–81, 2009.

[165] Z. A. Chen, "A new two-parameter lifetime distribution with bathtub shape or increasing failure rate function," *Stat. Probab. Lett.*, vol. 49, pp. 155–161, 2000.

[166] M. Bebbington, C.-D. Lai, and R. c. Zitikisb, "A flexible Weibull extension," *Reliab. Eng. Syst. Safe*, vol. 92, pp. 719–726, 2007.

[167] W. Barreto-Souza, A. Santos, and G. M. Cordeiro, "The beta generalized exponential distribution," *J. Stat. Comput. Simul.*, vol. 80, pp. 159–172, 2010.

[168] S. Nadarajah, G. M. Cordeiro, and E. M. M. Ortega, "General results for the Kumaraswamy-g distribution," *J. Stat. Comput. Simul.*, vol. 82, pp. 951–979, 2012.

[169] P. F. Paranaíba, E. M. M. Ortega, G. M. Cordeiro, and M. A. R. de Pascoa, "The Kumaraswamy Burr XII distribution: theory and practice," *J. Stat. Comput. Simul.*, vol. 83, pp. 2117–2143, 2013.

[170] G. M. Cordeiro, S. Nadarajah, and E. Ortega, "The Kumaraswamy Gumbel distribution," *Stat. Method Appl.*, vol. 21, pp. 139–168, 2012.

[171] M. Xie and C. D. Lai, "Reliability analysis using an additive Weibull model with bathtub-shaped failure rate function," *Reliab. Eng. Syst. Safe*, vol. 52, pp. 87–93, 1995.

[172] E. M. Wright, "The asymptotic expansion of the generalized hypergeometric function," *J. London Math. Soc.*, vol. 10, pp. 286–293, 1935.

[173] W. J. Zimmer, J. B. Keats, and F. K. Wang, "The Burr XII distribution in reliability analysis," *J. Qual. Technol.*, vol. 30, pp. 386–394, 1998.

[174] Q. Shao, "Notes on maximum likelihood estimation for the three-parameter Burr XII distribution," *Comput. Stat. Data An.*, vol. 45, pp. 675–687, 2004.

[175] Q. Shao, H. Wong, and J. Xia, "Models for extremes using the extended three parameter Burr XII system with application to flood frequency analysis." *Hydrolog. Sci. J.*, vol. 49, pp. 685–702, 2004.

[176] A. A. Soliman, "Estimation of parameters of life from progressively censored data using Burr-XII model," *IEEE Trans. Rel.*, vol. 54, pp. 34–42, 2005.

[177] G. O. Silva, E. M. M. Ortega, V. G. Cancho, and M. L. Barreto, "Log-Burr XII regression models with censored data," *Comput. Stat. Data An.*, vol. 52, pp. 3820–3842, 2008.

[178] R. L. Smith and J. C. Naylor, "A comparison of maximum likelihood and bayesian estimators for the three- parameter Weibull distribution," *J. R. Stat. Soc. Series C Appl. Stat.*, vol. 36, pp. 358–369, 1987.

[179] D. F. Andrews and A. M. Herzberg, *Data. A Collection of Problems from Many Fields for the Student and Research Worker.* New York: Springer-Verlag, 1985.

[180] K. Cooray and M. Ananda, "A generalization of the half-normal distribution with applications to lifetime data," *Commun. Stat.-Theory Methods*, vol. 37, pp. 1323–1337, 2008.

[181] A. N. F. Silva, "Estudo evolutivo da crianças expostas ao HIV e notificadas pelo núcleo de vigilância epidemiológica do HCFMRP-USP," Ph.D. dissertation, Universidade de São Paulo Faculdade de Medicina de Riberão Preto, 2004.

[182] G. S. C. Perdoná, "Modelos para dados de riscos aplicados à análise de sobrevivência," Ph.D. dissertation, Doutorado em Ciências da Computação e Matemática Computacional. Universidade de São Paulo, USP, Brasil., 2006.

[183] M. K. Cowles and B. P. Carlin, "Markov Chain Monte-Carlo convergence diagnostics: a comparative study," *J. Am. Stat. Assoc.*, vol. 91, pp. 883–904, 1996.

[184] A. Gelman and D. B. Rubin, "Inference from iterative simulation using multiple sequences," *Stat. Sci.*, vol. 7, pp. 457–472, 1992.

[185] S. Kotz and S. Nadarajah, *Extreme Value Distributions: Theory and Applications.* Imperial College Press, London, 2000.

[186] B. Efron, "Bootstrap methods: Another look at the jackknife," *Ann. Stat.*, vol. 7, pp. 1–26, 1979.

[187] B. Efron and R. J. Tibshirani, *An Introduction to the Bootstrap.* Taylor & Francis, 1994.

[188] T. J. DiCiccio and B. Efron, "Bootstrap confidence intervals," *Stat. Sci.*, vol. 11, pp. 189–228, 1996.

[189] N. L. R. Andrade, R. M. P. Moura, and A. Silveira, "Determinação da Q7,10 para o Rio Cuiabá, Mato Grosso, Brasil e comparação com a vazão regularizada após a implantação do reservatório de aproveitamento múltiplo de manso." in XXIV Congresso Brasileiro de Engenharia Sanitária e Ambiental, Belo Horizonte, Minas Gerais Brasil., 2007.

[190] M. R. Kazemi, H. Haghbin, and J. Behboodian, "Another generalization of the skew normal distribution," *World Appl. Sci. J.*, vol. 12, pp. 1034–1039, 2011.

[191] T. V. F. Santana, E. M. M. Ortega, G. M. Cordeiro, and G. O. Silva, "The Kumaraswamy-log-logistic distribution," *J. Stat. Theory Appl.*, vol. 11, pp. 265–291, 2012.

[192] H. Saulo, J. Leão, and M. Bourguignon, "The Kumaraswamy Birnbaum-Saunders distribution," *J. Stat. Theory Pract.*, vol. 6, pp. 745–759, 2012.

[193] G. M. Cordeiro, R. R. Pescim, and E. M. M. Ortega, "The Kumaraswamy generalized half-normal distribution for skewed positive data," *J. Data Sci.*, vol. 10, pp. 195–224, 2012.

[194] G. M. Cordeiro, E. M. M. Ortega, and G. O. Silva, "The Kumaraswamy modified Weibull distribution: theory and applications," *J. Stat. Comput. Simul.*, vol. 84, pp. 1387–1411, 2014.

[195] M. A. Correa, D. A. Nogueira, and E. B. Ferreira, "Kumaraswamy normal and Azzalini's skew normal modeling asymmetry," *Sigmae*, vol. 1, pp. 65–83, 2012.

[196] M. Bourguignon, R. B. Silva, L. M. Zea, and G. M. Cordeiro, "The Kumaraswamy Pareto distribution," *J. Stat. Theory Appl.*, vol. 12, pp. 129–144, 2013.

[197] S. Nadarajah and S. Eljabri, "The Kumaraswamy GP distribution," *J. Data Sci.*, vol. 11, pp. 739–766, 2013.

[198] I. Elbatal, "Kumaraswamy generalized linear failure rate distribution," *Indian J. Comput. Appl. Math.*, vol. 1, pp. 61–78, 2013.

[199] T. M. Shams, "The Kumaraswamy-generalized Lomax distribution," *Middle East J. Sci. Res.*, vol. 17, pp. 6–41, 2013.

[200] A. E. Gomes, C. Q. da Silva, G. M. Cordeiro, and E. M. M. Ortega, "A new lifetime model: the Kumaraswamy generalized Rayleigh distribution," *J. Stat. Comput. Simul.*, vol. 84, pp. 290–309, 2014.

[201] K. Zografos and N. Balakrishnan, "On families of beta- and generalized gamma-generated distributions and associated inference," *Stat. Methodol.*, vol. 6, pp. 344–362, 2009.

[202] S. Nadarajah, G. Cordeiro, and E. Ortega, "The Zografos-Balakrishnan-G family of distributions: Mathematical properties and applications," *Commun. Stat.-Theory Methods*, vol. 44, pp. 186–215, 2015.

[203] E. W. Stacy, "A generalization of the gamma distribution," *Ann. Stat.*, vol. 33, pp. 1187–1192, 1962.

[204] P. Flajonet and A. Odlyzko, "Singularity analysis of generating function," *SIAM J. Discrete Math.*, vol. 3, pp. 216–240, 1990.

[205] M. M. Ristić and N. Balakrishnan, "The gamma-exponentiated exponential distribution," *J. Stat. Comput. Simul.*, vol. 82, pp. 1191–1206, 2012.

[206] M. W. A. Ramos, G. M. Cordeiro, P. R. D. Marinho, C. R. B. Dias, and G. G. Hamedani, "The Zografos-Balakrishnan log-logistic distribution: Properties and applications," *J. Stat. Theory Appl.*, vol. 12, pp. 225–244, 2013.

[207] K. Adamidis and S. Loukas, "A lifetime distribution with decreasing failure rate," *Stat. Probab. Lett.*, vol. 39, pp. 35 – 42, 1998.

[208] T. K. Boehme and R. E. Powell, "Positive linear operators generated by analytic functions," *SIAM J. Appl. Math.*, vol. 16, pp. 510–519, 1968.

[209] S. Ostrovska, "Positive linear operators generated by analytic functions," in *Proceedings of the Indian Academy of Sciences*, vol. 117, 2007, pp. 485–493.

[210] C. Kuş, "A new lifetime distribution," *Comput. Stat. Data An.*, vol. 51, pp. 4497–4509, 2007.

[211] M. H. Chen, J. G. Ibrahim, and D. Sinha, "A new bayesian model for survival data with a surviving fraction," *J. Am. Stat. Assoc.*, vol. 94, pp. 909–919, 1999.

[212] F. Cooner, S. Banerjee, B. P. Carlin, and D. Sinha, "Flexible cure rate modeling under latent activation schemes," *J. Am. Stat. Assoc.*, vol. 102, pp. 560–572, 2007.

[213] R. Tahmasbi and S. Rezaei, "A two-parameter lifetime distribution with decreasing failure rate," *Comput. Stat. Data An.*, vol. 52, pp. 3889–3901, 2008.

[214] W. Lu and D. Shi, "A new compounding life distribution: the Weibull-Poisson distribution," *J. Appl. Stat.*, vol. 39, pp. 21–38, 2012.

[215] A. L. Morais and W. Barreto-Souza, "A compound class of Weibull and power series distributions," *Comput. Stat. Data An.*, vol. 55, pp. 1410–1425, 2011.

[216] V. G. Cancho, F. Louzada-Neto, and G. D. Barriga, "The Poisson-exponential lifetime distribution," *Comput. Stat. Data An.*, vol. 55, pp. 677–686, 2011.

[217] V. G. Cancho, F. Louzada-Neto, and G. D. C. Barriga, "The geometric Birnbaum Saunders regression model with cure rate," *J. Stat. Plan. Inference*, vol. 142, pp. 993–1000, 2012.

[218] W. Barreto-Souza and H. S. Bakouch, "A new lifetime model with decreasing failure rate," *Stat.*, vol. 47, pp. 465–476, 2013.

[219] R. B. Silva, M. Bourguignon, D. C. R. B., and G. M. Cordeiro, "The compound family of extended Weibull power series distributions," *Comput. Stat. Data An.*, vol. 58, pp. 352–367, 2013.

[220] M. Gurvich, A. DiBenedetto, and S. Ranade, "A new statistical distribution for characterizing the random strength of brittle materials," *J. Mater. Sci.*, vol. 32, pp. 2559–2564, 1997.

[221] M. Bourguignon, R. B. Silva, and G. M. Cordeiro, "A new class of fatigue life distributions," *J. Stat. Comput. Simul.*, vol. 84, pp. 2619–2635, 2013.

[222] R. B. Silva and G. M. Cordeiro, "The Burr XII power series distributions: A new compounding family," *Brazilian J. Probab. Stat.*, vol. 29, pp. 565–589, 2015.

[223] N. L. Johnson, S. Kotz, and N. Balakrishnan, *Continuous Univariate Distributions.* John Wiley and Sons, New York, 1994, vol. 1.

[224] J. W. S. Rayleigh, "On the resultant of a large number of vibrations of the same pitch and of arbitrary phase," *Philos. Mag.*, vol. 10, pp. 73–78, 1880.

[225] J. S. White, "The moments of log-Weibull order statistics," *Technometrics*, vol. 11, pp. 373–386, 1969.

[226] K. K. Phani, "A new modified Weibull distribution function," *J. Am. Ceram. Soc.*, vol. 70, pp. 182–184, 1987.

[227] J. A. Kies, *The strength of glass.* Washington D.C. Naval Research Lab, 1958.

[228] R. Smith and L. Bain, "An exponential power life-testing distribution," *Commun. Stat.-Theory Methods*, vol. 4, pp. 469–481, 1975.

[229] H. Pham, "A bathtub-shaped hazard rate function with applications to system safety," *Int. J. Reliab. Appl.*, vol. 3, pp. 1–16, 2002.

[230] F. Louzada, M. Roman, and V. G. Cancho, "The complementary exponential geometric distribution: Model, properties, and a comparison with its counterpart," *Comput. Stat. Data An.*, vol. 55, pp. 2516–2524, 2011.

[231] G. M. Cordeiro, J. Rodrigues, and M. de Castro, "The exponential COM-Poisson distribution," *Stat. Pap.*, vol. 53, pp. 653–664, 2012.

[232] J. D. Flores, P. Borges, V. Cancho, and F. Louzada, "The complementary exponential power series distributions," *Brazilian J. Probab. Stat.*, vol. 4, pp. 565–584, 2013.

[233] W. Barreto-Souza and R. Silva, "A likelihood ratio test to discriminate exponential–Poisson and gamma distributions," *J. Stat. Comput. Simul.*, vol. 85, pp. 802–823, 2015.

[234] W. Barreto-Souza, A. L. Morais, and G. M. Cordeiro, "The Weibull-geometric distribution," *J. Stat. Comput. Simul.*, vol. 81, pp. 645–657, 2010.

[235] H. M. Barakat and Y. H. Abdelkader, "Computing the moments of order statistics from nonidentical random variables," *Stat. Method Appl.*, vol. 13, pp. 13–24, 2004.

[236] D. R. Cox and D. V. Hinkley, *Theoretical Statistics.* Chapman and Hall, London., 1974.

[237] A. Alzaatreh, F. Famoye, and C. Lee, "A new method for generating families of continuous distributions," *Metron*, vol. 71, pp. 63–79, 2013.

[238] M. Santos-Neto, M. Bourguignon, L. M. Zea, A. D. C. Nascimento, and G. Cordeiro, "The Marshall-Olkin extended Weibull family of distributions," *J. Stat. Distrib. Appl.*, vol. 1, no. 9, pp. 1–9, 2014.

[239] P. Sankaran and K. Jayakumar, "On proportional odds model," *Stat. Pap.*, vol. 49, pp. 779–789, 2008.

[240] K. Jayakumar and T. Mathew, "On a generalization to Marshall-Olkin scheme and its application to Burr type XII distribution," *Stat. Pap.*, vol. 49, pp. 421–439, 2008.

[241] J. McDonald, "Some generalized functions for the size distribution of income," *Econometrica*, vol. 52, pp. 647–63, 1984.

[242] A. Alzaatreh, F. Famoye, and C. Lee, "Weibull-Pareto distribution and its applications," *Commun. Stat.-Theory Methods*, vol. 42, pp. 1673–1691, 2013.

[243] H. Torabi and N. H. Montazari, "The gamma-uniform distribution and its application," *Kybernetika*, vol. 48, pp. 16–30, 2012.

[244] M. Amini, S. M. T. K. MirMostafaee, and J. Ahmadi, "Log-gamma-generated families of distributions," *Stat.*, vol. 48, pp. 913–932, 2014.

[245] H. Torabi and N. H. Montazari, "The logistic-uniform distribution and its application," *Commun. Stat.-Simul. Comput.*, vol. 43, pp. 2551–2569, 2014.

[246] G. M. Cordeiro, E. M. M. Ortega, and D. C. C. da Cunha, "The exponentiated generalized class of distributions," *J. Data Sci.*, vol. 11, pp. 1–27, 2013.

[247] A. Alzaghal, C. Lee, and F. Famoye, "Exponentiated T-X family of distributions with some applications," *Int. J. Stat. Probab.*, vol. 2, pp. 31–49, 2013.

[248] M. Bourguignon, R. B. Silva, and G. M. Cordeiro, "The Weibull–G family of probability distributions," *J. Data Sci.*, vol. 12, pp. 53–68, 2014.

[249] G. M. Cordeiro, M. Alizadeh, and E. M. M. Ortega, "The exponentiated half-logistic family of distributions: Properties and applications," *J. Probab. Stat.*, vol. 2014, pp. 1–21, 2014.

[250] M. H. Tahir, G. M. Cordeiro, A. Alzaatreh, M. Zubair, M. Mansoor, and M. Alizadeh, "The logistic-X family of distributions and its applications," *Commun. Stat.-Theory Methods*, vol. 45, pp. 7326–7349, 2016.

[251] A. Alzaatreh, F. Famoye, and C. Lee, "Gamma-Pareto distribution and its applications," *J. Mod. Appl. Stat. Methods*, vol. 11, pp. 78–94, 2012.

[252] ——, "The gamma normal distribution: Properties and applications," *Comput. Stat. Data An.*, vol. 69, pp. 67–80, 2014.

[253] M. H. Tahir, M. Zubair, M. Mansoor, G. M. Cordeiro, and M. Alizadeh, "A new Weibull-G family of distributions," *Hacet. J. Math. Stat.*, vol. 45, pp. 29–647, 2016.

[254] G. Cordeiro, E. Ortega, B. Popovic, and R. Pescim, "The Lomax generator of distributions: Properties, minification process and regression model," *Appl. Math. Compt.*, vol. 247, pp. 465–486, 2014.

SUBJECT INDEX

A

Accelerated failure time model 8, 10
Adequacy Model script 174
AIC and BIC statistics 51, 95, 96, 97, 196, 197
  for expenditure data 197
AIDS data 132, 133
Akaike information criterion (AIC) 24, 121, 122, 131, 133, 134, 169, 170, 174, 175, 198
Algebraic 13, 72, 121
  calculations 121
  expansions 72
  manipulations, minor 13
Analysis 1, 2, 8, 23, 53, 56, 83, 123, 127, 129
  diagnostic 8
  dynamic 129
  methods 2
  flood frequency 123
  statistical 1
  statistical data 53
  survival 2, 23, 56, 83, 127
Analytical intractability 143
Applications 2, 137, 178, 180
  associated 2
  complex 137
  fifty 137
  industrial 178, 180
Applied mathematics and statistics 71
Approximate confidence intervals 129, 141, 173
Asymptotes 99, 105, 114, 116, 138, 153, 155
  and shapes 114, 153
Asymptotic 37, 45, 50, 55, 57, 85, 128, 130, 141, 183
  distributions 37, 45, 55, 57, 85, 128, 141, 183
  normality 50
  theory, first-order 130

B

Ball bearings, deep-groove 183
Baseline 8, 60, 65, 66, 80, 85, 99, 100, 106, 172, 180, 188, 190, 200, 201, 203
  arbitrary 201
  cdf, arbitrary 106
  density function 180
  hazard function 8
  laws 8
  Weibull distribution 172
Baseline distributions 8, 27, 65, 66, 70, 115, 178, 180, 183, 188, 191, 198, 200, 201
  adopted 178
  arbitrary 115
  exponentiated 188, 191
Baseline models 26, 27, 59, 100, 189, 200
  exponentiated 189
Bathtub hazard rates 31, 33
  upside-down 31
Bayesian 24, 29, 56, 95, 129, 135, 143, 146, 147, 198
  analysis 129, 135, 143, 146
  and L-moment methods 29
  approach 143
  information criterion (BIC) 24, 95, 147, 198
  updating 129
BBS and BS distributions 95
BCa 142, 143, 145
  bootstrap interval 143
  confidence intervals 145
  method 142
Bernoulli distribution 27
Beta-Birnbaum-Saunders (BBS) 57, 87, 94, 97, 98, 121, 145
Beta distribution 54, 57, 59, 60, 74, 81, 100, 201
  generalized 201
  inverted 57, 59
Beta exponentiated exponential (BEE) 111
Beta generalized exponential (BGE) 31, 56, 87, 88, 98

www.ingramcontent.com/pod-product-compliance
Lightning Source LLC
Chambersburg PA
CBHW050826220326

41598CB00006B/323